THE SUN:
OUR FUTURE ENERGY SOURCE

THE SUN:
OUR FUTURE ENERGY SOURCE

SECOND EDITION

DAVID K. McDANIELS
University of Oregon

JOHN WILEY & SONS

New York • Chichester • Brisbane • Toronto • Singapore

Library of Congress Cataloging in Publication Data:

McDaniels, David K., 1929–
    The sun, our future energy source.

    Includes bibliographies and index.
    1. Solar energy. I. Title.
TJ810.M17   1984   621.47   83-16758
ISBN 0-471-86859-0

Printed in the United States of America

10 9 8 7 6 5 4 3 2

preface
to the
first edition

Solar energy is certainly not a new concept. It has long been realized that, despite the low energy density of the incoming insolation, solar radiation has a large potential as an energy source. Although the possibility of electric power generation from solar radiation is very promising for the future, it is the application of solar energy techniques to the heating and cooling of buildings that lies immediately within the scope of present day state-of-the-art technology. The solar energy falling upon the roof of a typical residence is roughly 10 times as great as the annual heat demand for that house. Since almost 25 percent of our nation's energy consumption goes into space heating, space cooling, and water heating, this means that the rapid development of the solar heating (and cooling) industry could make a significant impact upon this country's energy budget by the turn of the century.

There is a substantial base of research upon which the application of solar energy to space heating and cooling can be built. Between 1940 and 1970 more than 25 houses and laboratory installations were partially heated with solar energy. The reason it has taken so long to implement the widespread use of solar energy is that while the solar fuel is free, a considerable capital investment is needed to harness it. The availability of cheap oil, electricity, and natural gas has seriously impeded solar energy applications in the past. Only recently, because of the escalation in fossil fuel costs, has solar energy become economically competitive for space heating. The economic position of solar energy for space and water heating and for air conditioning should continue to improve in the next few years as the cost of conventional energy rises.

This book originated from lecture notes for a one-quarter introductory course on solar energy taught by the author at the University of Oregon. The course is one of several "minicourses" offered by the Physics Department to provide a liberal arts student with a broad background in scientific ideas and methods while dealing with topics that are of real interest to a wide spectrum of nonscience majors. Our success with these courses has more than justified the work required to prepare them. The enrollment in the course on solar energy increased from 25 to 450 students over a period of six years, reflecting the rapid increase of interest in this topic.

The principal aim of this text is to help the reader gain an understanding of the ways in which useful energy can be obtained from the incoming solar radiation. Thus, the major portion of this textbook covers topics such as solar radiation, flat-plate collectors, active and passive solar heating systems, solar thermal electric generating schemes, and photovoltaic devices. In order to round out the course and provide the requisite background, the first three chapters discuss topics related to the overall energy problem, such as the energy crisis, fossil fuel reserves, and nuclear fission. Some of this introductory material can be readily left out of the course, if desired, without seriously affecting the student's understanding of future developments.

In keeping with the introductory nature of the course, the presentation has been kept as nonmathematical as possible. The emphasis is entirely upon developing a qualitative understanding of the subject. Numerical examples are occasionally worked out to illustrate the principles involved, however students are not expected to be able to work out detailed calculations themselves. The problems at the end of each chapter were chosen to span a large range of difficulty. Some of the multiple-choice questions are exceedingly easy, included only to check that the reader is aware of the most basic facts. I hope the others will cause the interested reader to seriously think about the subject matter under discussion.

The system of units used in this book is the MKS system, which is essentially the SI system of units, with certain modifications. In this metric system the basic units of length, mass, and time are the meter, kilogram, and second. The corresponding metric unit of energy is the joule. Occasionally, English units have been used when it was felt that current practice demanded their usage.

Most people seriously interested in solar energy developments are also deeply concerned about energy conservation, recycling of valuable mineral resources, and pollution of our environment. Because of this concern, the book begins with a chapter concerning the question of energy growth and resource usage in a finite world. Chapter 2 introduces the student to the concepts and uses of energy. The resource potential of fossil fuels, geothermal energy, and nuclear fission is assessed in Chapter 3. These chapters are intended to provide the necessary background about the present energy crisis, about the concept of energy, and about our ability to meet this crisis with nonrenewable energy sources.

The coverage of solar energy begins in Chapter 4 with a short, historical review and an assessment of the economic viability of solar energy for space heating as compared with our present energy alternatives. This discussion is followed by a slight diversion into cosmology in which the nature of the universe and the formation of the sun are described. Purists may object to the inclusion of this material, but our experience indicates a large student interest in how the sun was formed, how it compares with other stars, and how it produces the energy that we receive each day.

A rather extensive coverage of solar radiation is presented in Chapter 6. The concepts of wave motion, and light, the breakdown of the incident solar energy into direct and scattered components, and radiation monitoring instruments are among the topics examined. Also included is a brief discussion of blackbody radiation, the connection between emissivity and absorptivity, and Rayleigh scattering.

The last five chapters discuss the primary solar energy applications: space heating and cooling and solar electric power generation. Chapter 7 begins with a simplified description of how flat-plate collectors work. The basic concepts relating to efficient collection design, heat loss analysis, details of construction, and selective surfaces are discussed. The following chapter integrates the flat-plate collector into the overall heating system. A solar heating system requires a number of unique features by contrast with conventional systems. Topics covered include heat-exchange fluids, heat storage, delivery and control of the collected solar heat, solar cooling, and the use of reflector enhancement. A discussion of active and passive solar houses is contained in Chapter 9. While active solar systems have been emphasized in this book in order to better explain the physical principles involved, it is quite likely that passive approaches will prove to be the most economic way to utilize solar energy for space heating.

While technically feasible right now, the economic development of solar electric power plants will require a massive and sustained program of research. The most promising ways to do this using thermally driven generating devices are reviewed in Chapter 10. All of these approaches involve the use of a heat engine of some type. Hence, the chapter begins with an elementary discussion of heat and thermodynamics and the restrictions upon operating efficiency that result because of the second law of thermodynamics. The solar electric approaches covered include the use of a large field of tracking mirrors to reflect the sunlight upon a central receiver, intermediate-temperature solar energy collection using a large array of linear focusing collectors, and the generation of electricity using ocean temperature differences.

Another promising way to convert solar energy to electricity is with the aid of solar cells. These devices have been used in the space effort since 1958 and, in their early stage of development, the cost was of the order of $10 million per peak kilowatt. Improvements in manufacturing techniques have reduced this to less than $15,000 per peak kilowatt in 1977, which illustrates the considerable progress made in this area. However, at this price, solar cells are still prohibitively expensive for large-scale use. Basic principles of photovoltaic devices as well as promising lines of research and development are discussed in Chapter 11.

It is impossible to acknowledge fully the many sources of information that I have utilized in writing this book. A short, annotated bibliography of some of the more important and useful references are listed at the end of each chapter. Special thanks are owed to Dr. W. A. Shurcliff for contributing numerous, useful suggestions. I gratefully acknowledge many useful discussions and assistance from my colleagues at Oregon: Professors D. H. Lowndes and John Reynolds, and Steve Baker and Dan Kaehn. A great deal of stimulating and rewarding feedback was provided by the questions and remarks from the students in my classes. Their input played an important role in shaping the coverage and level of this book. Finally, this textbook could not have been written without the extensive secretarial assistance provided by Sandy Hill and the extensive secretarial and editorial assistance of my wife, Pat.

David K. McDaniels

preface

Rapid advances in technology and major changes in emphasis in solar energy have occurred during the past few years. Examples of the former include the development of the 10-megawatt electric solar tower facility at Barstow, California and the spectacular increase in efficiency of amorphous silicon solar cells. An example of the latter is the shift toward passive solar and away from active systems for space heating applications.

A major change in this edition is the addition of an entire chapter on passive solar applications. The chapter on active solar systems is rewritten stressing solar hot water applications. The last two chapters on solar electricity are updated to reflect the progress made over the past five years. Although the basic outline of the book is retained, some of the chapter headings are revised to better reflect the modifications in content. An important pedagogical change is the inclusion of answers to the end-of-chapter questions. In addition, many new questions have been added.

Further details about the changes in various chapters are worth mentioning. Chapter 2 has been extensively modified and now includes the section on thermodynamics. Chapter 4 is now completely devoted to solar history. The material describing a few classical examples of active solar houses has been moved to this earlier chapter. In Chapter 6 the discussion of the direct and diffuse components of solar radiation is simplified.

A new chapter (Chapter 7) is devoted entirely to passive solar topics. This material is placed just after the coverage of solar radiation and before the discussion of active space and water heating. This emphasis reflects the tremendous growth in interest in this area over the past few years. A very complete treatment of direct-gain and thermal storage wall applications is presented. Also included are treatments of new approaches such as double-envelope and superinsulated houses.

The coverage in Chapter 8 has been expanded from that in the first edition to include an elementary discussion of concentrating devices. This is followed in the next chapter (on active solar systems) by a greatly expanded treatment of solar hot water systems. Solar hot water applications are probably the most practical and economic solar application at this time.

The changes in the chapter on solar thermal electricity involve an expanded treatment of the solar tower concept and the removal of the section on thermodynamics to Chapter 2. The last chapter describing how electricity is obtained with solar cells is entirely rewritten. The section on thermionic devices has been removed, and the coverage of solar cell fundamentals is extensively modified. The lure and promise of the new area of amorphous silicon is discussed in some detail.

I gratefully acknowledge many useful discussions with my colleague Frank Vignola. I also express my gratitude to Dave Fong, who contributed two new figures to the chapter on solar radiation and to my colleagues Kwangjai Park and Joel W. McClure, who made many useful suggestions for improvement. Finally, this revision would not have been possible without the support, understanding, and encouragement of my wife, Pat.

David K. McDaniels

contents

THE SUN:
OUR FUTURE ENERGY SOURCE

1 our finite world

Nearly every aspect of the modern world is built upon the availability of abundant and relatively inexpensive energy. Our industrial development, our abundance of material goods, our food supply and agricultural practices, nearly all the conveniences of our everyday lives depend on a ready supply of energy. The extent to which all facets of our society have become dependent on energy can be illustrated by looking at the changes in agricultural practices over the last 60 years. Figure 1-1 depicts the rapid development of mechanized farming since 1900. The horse is no longer used as a prime mover.

The source of most of our energy has been our nonrenewable fossil fuels, oil, coal, and natural gas. This is unfortunate from a long-term point of view as these precious resources that took millions of years for nature to produce are being depleted in a few hundred years. This rapid use of our nonrenewable resources is being matched by an equally voracious consumption of our other precious nonrenewable natural resources, such as minerals like magnesium, manganese, and titanium.

Our present overdependence on nonrenewable natural resources could lead to catastrophic consequences. There is real evidence that human beings have outstripped the capacity of this planet to sustain present life-styles. Our inescapable problem is that we continue to consume nonrenewable resources at a rate that will give rise to serious problems in the near future. It is now time for us to understand that a large fraction of the current world problems are not just a result of

1

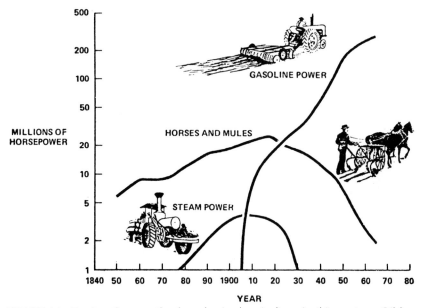

FIGURE 1-1 Depicts the growth of mechanized agriculture in this century. Oil has essentially replaced the horse as the prime mover on farms. (Courtesy of Harrison Brown. Reproduced, with permission, from the *Annual Review of Energy*, Vol. 1. Copyright © 1976 by Annual Reviews Inc.)

economic or political failures but, instead, are related to the world's diminishing natural resources. The solution lies in conservation and a change from our reliance on nonrenewable resources to solar energy and other renewable resources.

A. GROWTH OF ENERGY USAGE

The United States has enjoyed an abundance of raw energy resources including wood, coal, oil, hydropower, natural gas, and uranium. But our consumption of these energy resources has been reckless. With 6% of the world's population, the U.S. share of total world energy use is roughly 30%. (The U.S. energy usage was about 8×10^{19} joules (J) in 1980.) Our per capita energy consumption is about 4 times that of Western Europe and 20 times that of Communist Asia. Over the course of the last five decades the use of energy in the United States in all forms has been doubling every 15 years. Until recently electrical energy production has been doubling every 9 to 10 years. This growth in energy usage has not varied simply in response to a corresponding growth of population. Energy usage in the United States, in particular, and in developed nations generally, has grown more rapidly than population, because of the simultaneously rising material standard of living and because of the continuous change to more energy-intensive methods of providing that standard of living. This is illustrated in Figure 1-2, which shows the increase in energy and population for the United States in the twentieth century.

The earth is a system with limited supplies of land, air, water, vegetation, and minerals. The rapid increase in energy use characteristic of the past 50 to 100 years cannot continue indefinitely in a finite system. At some point even the aesthetic damage to our lives from too many generating stations, too many transmission lines, too many factories, and the like will become intolerable. One serious aspect of the increasing rate of consumption of our energy resources arises from the large amounts of waste heat given off. This waste heat could threaten the earth's environmental balance if our energy growth continues unheeded. For example, it has been estimated that a 2°F rise in the earth's temperature could melt the Arctic Ocean ice pack. As a more immediate problem, thermal pollution of some of our rivers has noticeably changed their customary ecological balance. Urban climatic conditions could also be affected. It has been estimated that if the present rate of increase of energy use in the Los Angeles area continues, a 5 to 7° rise in the ambient temperature will be experienced in the next 20 years.

The type of growth in energy resource consumption that we have witnessed throughout most of this century is exponential growth; growth characterized by the existence of a *doubling time*, during which the energy consumed doubles, then redoubles again in another time

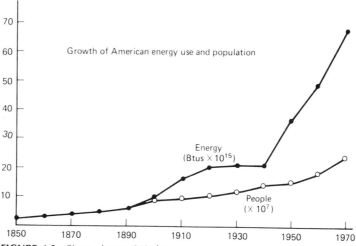

FIGURE 1-2   Since about 1940 the growth in consumption of our energy resources has increased more rapidly than the population.

interval equal to the doubling time, and so on. This exponential growth behavior is illustrated in Figure 1-3. As an example of exponential growth pushed to ridiculous extremes, consider that the amount of energy available from fusion of the deuterium in the oceans is sufficient (provided that controlled deuterium fusion reactors become available) to last for more than a billion years at our 1980 rate of consumption of energy. However, at our present doubling rate of 15 years, this practically infinite resource of energy would be consumed in 450 years. Of course, many other problems would arise long before this occurs.

Growth characterized by the existence of a doubling time eventually outstrips all of the finite resources in nature and never continues unchecked in a finite system. The only naturally occurring instances of exponential increase are inevitably either transient (temporary) phenomena or are terminated by destruction of the system in which the growth takes place. Cancerous cell growth is one such naturally occurring example. The exponential increase of an animal population to a point where overgrazing and starvation of the entire population

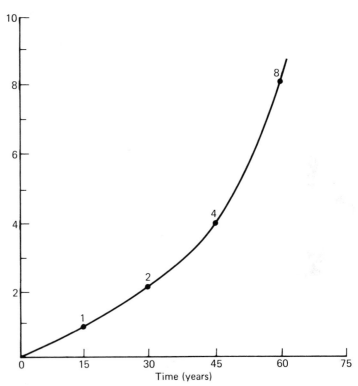

FIGURE 1-3   Graph illustrating the growth of energy usage with time assuming that the amount of energy consumed doubles every 15 years. This type of growth is called exponential growth.

results is another. The balance that can be reached between two populations of predator and prey animals shows how nature normally prevents exponential growth by limiting the increase in any one animal population to a size that can live in harmony with its surroundings. This type of balance, in physics, is called a "steady-state" situation.

Only recently have doubts been raised about the legitimacy of our present and future energy needs. Estimates of future energy needs were (and still are) based largely on projections of existing rates of energy consumption; these projections are most often simply extrapolations into the future of present growth rates. Previously, no one questioned the right of economic growth and resultant exponential increase in energy consumption. Growth in terms of population, consumption of goods, improved technology, use of energy, and pollution has been an important factor in determining the character of our society. With the increasing concern for the environment and the growing realization that oil supplies are dwindling, the wisdom of this pattern of behavior is now being questioned.

B. THE ENERGY CRISIS

Because our present energy supplies are both finite and nonrenewable, our "energy splurge" appears to be coming to an end. Warning signs in the form of oil shortages and blackouts have been given. In the winter of 1971 Britain faced a serious lack of electric power, because of a nationwide strike of coal miners that lasted for a period of two months. Without electricity, activity in a modern nation soon lessens. The available electric power was rationed in some areas during blackouts for as little as 9 hours a day. For several weeks the inhabitants of Britain were made starkly aware of the fact that electricity has become essential to modern life. At home the domestic oil shortage of 1973 brought an awareness of our critical dependence on depletable energy resources.

The heart of the present energy problem is the fact that the United States relies on oil for almost one-half of its supply of energy. Table 1-1 shows a comparison betweeen the U.S. sources of energy for 1973 and 1979. A detailed presentation of the U.S. resources is given in Chapter 3, but the inescapable fact is that although one-half of our energy consumption comes from oil, our own production is now going down, with only a brief respite expected from the advent of Alaskan supplies. Generous estimates of how much oil might ultimately be extracted from the North Slope Fields indicate that they will only satisfy present U.S. petroleum needs for about a decade at most. Similarly, Table 1-1 illustrates that our dependence on oil did not significantly decrease

TABLE 1-1
Origin of the Energy Uses in the United States

	PERCENTAGE OF TOTAL	
SOURCE	1973	1979
Oil	47.7	46.9
Coal	18.3	18.9
Natural gas	31.4	24.9
Nuclear	1.3	6.0
Hydroelectric and renewables	1.3	3.3

between 1973 and 1979, despite much public rhetoric about the United States becoming energy independent.

There are certainly a number of domestic energy resources that can be used to compensate for the impending shortages. Substantial quantities of oil have been left behind in "exhausted" reservoirs. This country has great potential resources of oil shale. Our proved coal resources could last for several hundred years at the present consumption rate. If the breeder reactor proves feasible (and acceptable) our reserves of uranium could satisfy our needs for several thousand years. Also, the United States has an enormous potential resource of geothermal energy, if it can be satisfactorily utilized. Finally, solar energy represents an almost inexhaustible source of energy.

Where then lies the problem? Harrison Brown, Professor of Geochemistry at California Institute of Technology, aptly characterizes the situation by pointing out that we have become ensnared in an "energy trap," where inexpensive energy is taken for granted. According to Brown this has:

1. Caused us to take extremely low energy costs for granted and to accept them as the norm to be expected for a very long time in the future.

2. Deterred the development of a broadly based industrial expertise in the use of alternative energy sources.

3. Caused us to give low priority to research programs aimed at developing alternative energy supplies.

4. Prevented us from investing adequate amounts of capital in known technological processes for utilizing alternative energy sources safely on a substantial scale.

To this list we might add that this energy trap has hindered the development of adequate conservation measures.

Because energy is relatively cheap, it is used in many wasteful ways. The cost of the electricity used to produce common items such as paper, food, machinery, wood products, and transportation equipment is only 1 to 2% of total production cost. It is no wonder that only recently have steps been taken to decrease the amount of energy used by industry, for heating and cooling residences, and for transportation. It is not our purpose here to analyze the reasons why cost for energy is as low as it is. But certainly federal subsidies and relaxed environmental regulations have contributed. Will higher energy costs in the future seriously change the basic economic structure of the country? Naturally the answer to this question depends on just how far the cost of energy escalates in the future. In Western Europe the price of energy is at least twice that of the United States. But, this region still maintains a modern industrialized society.

There are also factors affecting energy production that have more to do with political and environmental decisions than with dwindling natural resources. For example, the federal government may have accentuated the natural gas shortage through its pricing policies. Antipollution measures have curbed the use of high-sulfur coal, while the government has provided insufficient support and incentive to the research necessary for the clean utilization of our plentiful coal reserves. Construction of new nuclear power stations has been opposed on grounds such as safety and radioactive waste disposal. Offshore exploration for oil and gas has also been slowed by environmental concerns. These environmental concerns are important, but it must be recognized that satisfying environmental requirements will raise the price of a unit of energy.

Provided that public policy can cope with the political and environmental considerations, there would appear to be adequate energy resources for both the near- and long-term future. Probably very few of us wants to go back to the "good old days," if it means completely giving up the energy-consuming trappings of our daily lives. Central heating, television, electric appliances, and other conveniences have become part of a way of life. But the era of cheap, unlimited power is coming to a close. No longer will we be able to design and construct residential and commercial buildings, industrial projects, and transportation systems in the most convenient manner possible, without regard for how they fit the overall energy use and conservation picture.

C. ENERGY CONSERVATION

What can be done to remedy our energy crisis? Clearly we need to continue development of energy resources other than petroleum. But this will take time. An immediate solution is to embark on a large-scale program to reduce the amount of energy that is wasted. Until recently energy conservation* was virtually ignored or dismissed. This view is typified by a report put out several years ago by the Chase Manhattan Bank that maintained the thesis that any major reduction in energy usage would harm both the nation's economy and its standard of living. The point of view of this report was that most energy resource consumption is for essential purposes; two-thirds goes into commercial business needs with most of the remainder providing for essential private needs. The conclusion was that although some minor uses of energy could be regarded as strictly nonessential, their elimination would not permit any significant savings.

Most informed people do not share this viewpoint. For example, the American Institute of Architects has pointed out that by 1990, more energy could be saved by energy conservation techniques in buildings than could be obtained from any of the following sources: domestic oil, conversion of shale rock to oil, Alaskan North Slope oil, domestic and imported natural gas, or nuclear energy. Individuals waste energy and there is much each of us can do to conserve. But it must be stressed that energy in this country has been wasted primarily because public and private policies have long promoted energy extravagance in the mistaken belief that this was the best way to achieve a prosperous economy.

The use of energy conservation techniques does not necessarily create a lower standard of living. The standard of living in Sweden is comparable to that in this country even though Swedes use less than one-half the energy per person used by Americans. According to Schipper (see references at the end of the chapter) 24% of the energy used in thermal power plants in Sweden is utilized for low-temperature process or space-heating purposes. In the United States this waste heat is almost entirely discarded to the environment. He also claims that an evaluation, taking into account the slightly smaller size of Swedish dwellings, the higher percentage of apartments, the higher efficiencies of the well-maintained apartment heating systems, and the Swedish climate, shows that space-heating requirements in Sweden are 30 to 50% lower per square foot of space in homes and commercial buildings than in the United States.

As a nation we have the potential to save almost a factor of 2 in our energy consumption through conservation without drastically curtailing our standard of living. Let us review briefly some of the possibilities. Major improvements can be made in the design and insulation of homes and commercial buildings. The most significant parameters for

---

* It must be emphasized that energy can neither be created nor destroyed. All that can be done is to transfer it from one form to another (for more detail, see Chapter 2). In this chapter we use the expression *energy conservation* to mean reducing usage and waste of our energy resources.

energy consumption by a residence are the ambient (outside) temperature and wind speed. Inadequate insulation and excessive infiltration
of outside air into homes increases the energy used for both heating and
cooling. One study of model homes in three different regions of the
country showed that additional insulation in walls and ceilings, proper
use of weather stripping, and foil insulation in floors would save an
average of 42% of the energy consumed for space heating. A similar
study of commercial buildings estimates that savings of up to 40%
could be obtained in these structures as well. The potential savings
from both could amount to more than 7% of our present national total
energy use.

Most buildings are overheated. Savings of 15 to 20% can readily be
obtained through the simple procedure of turning down thermostats.
This is easily shown by the following example. A typical home in the
Pacific Northwest might have an equilibrium temperature inside of 21
degrees Celsius (21°C) while the ambient temperature is 7°C. By
lowering the inside average temperature to 18°C, which could easily be
attained through operation at 21°C in the daytime and 15°C at night, a
savings of

$$1 - \frac{18 - 7}{21 - 7} = 1 - 0.79 = 0.21 \qquad (1\text{-}1)$$

or 21% can be attained. Operation at these temperatures may even be
better for one's health! More refined estimates of the heating season
savings across the country give roughly this result.

A window that is not receiving sunlight is responsible for a tremendous loss of heat in cold weather. Even a tightly weather-stripped
window of double glass loses 6 to 10 times as much heat as an equal
area of well-insulated wall. But in sunlight, the same window is often a
net gainer of heat. The sun's rays pass right through the glass and warm
the floors and walls inside. Each sun-heated floor or wall then becomes
in turn an emitter of infrared rays, but at a longer wavelength that will
not pass back out through the glass. The solar heat is effectively
trapped. A useful window design might appear as shown in Figure 1-4.
This is an example of the use of passive solar heating, which is
described more fully in Chapter 7. Another excellent passive approach
is the use of superinsulation to greatly reduce the building heat load.

Solar hot water heating is an excellent and economical way to
reduce usage of conventional energy resources. However, minimizing
the hot water load can increase the attractiveness of solar hot water
heating. Hot water conservation is one of those items that is not widely
recognized for its large potential for saving money. Yet, with a few
simple modifications, the domestic hot water load can be dramatically
reduced. A pressure relief valve installed just downsteam from the
main shutoff valve reduces the amount of water used; this can be

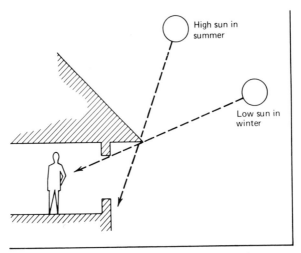

FIGURE 1-4 Proper overhang of eaves or awnings over window lets winter sunshine in to provide solar heating, yet screens out summer sun to help keep home cool.

supplemented with a low-flow showerhead. Another important conservation feature is to directly reduce energy usage by reducing the temperature of your hot water from 60 to 65°C (140 to 150°F) to 50°C (122°F). Additional gains can be made by adding an insulating jacket to the hot water tank. The most significant reduction in the hot water load can probably be obtained through the use of a *grey water recovery system*. This consists of a simple heat-exchange tank to take advantage of the waste heat from dishwashers, showers, and so on. Instead of heating the sewer pipes, some of this waste heat is used to warm up the incoming cold water to the hot water system. Some combination of the above techniques can easily reduce the hot water load by one-half. This represents a tremendous savings potential at the national level, since domestic hot water in the residential sector uses 3.1% of the nation's total energy use.

Improvements in the efficiency of utilization of energy can be made through the better design and implementation of some of our common household devices. The difference in energy efficiency among various models of common air conditioners is more than a factor of 3. Even better, one can use low-energy fans for cooling. Furnaces for space heating are often operated at only about one-half their stated efficiency because of poor maintenance and design. Electrically driven heat pumps could possibly double heat energy output. Frost-free refrigerators and freezers use almost twice the energy of those having a manual defrosting system. Fluorescent lights use only a quarter as much electricity as incandescent bulbs (although the reradiated energy from either type does go into useful space heating).

Improved design of modern high-rise buildings could lead to savings of the order of 5% of the national energy budget. Many large buildings are so poorly designed that they require air conditioning to take away the heat generated by the excessive use of lights. Architect Richard Stein believes that careful design—using reflective window glass, orienting the building properly with respect to the sun, reducing air exchange and interior illumination—could reduce the operating energy by one-half over conventional design. The 1454-ft Sears and Roebuck building in Chicago requires more electricity than the city of Rockford, Illinois, with 147,000 people. A basic energy-wasting factor in building design is the use of large glass expanses as typified by the Lever House in New York City. Glass transmits heat to the outside readily in winter and admits solar radiation in the summer, overtaxing the building's air-conditioning systems. The simplest solution is to substitute well-insulated walls for much of the glass area.

Overall savings of 10 to 20% could be made through improved industrial practices. The simplest way to achieve energy savings is to improve building insulation and reduce thermostats. Another way to obtain significant savings is to cascade high-temperature processes with lower-temperature demands. The whole idea of utilizing waste heat for useful purposes is almost unexplored in this country. Significant savings can also be made through the use of improved industrial technology. The aluminum industry offers an example of this. Older processing plants require about 20,000 kilowatt hours per ton (kWh/ton) while more efficient modern plants have cut this to about 9000 kWh/ton. It should also be noted that by recycling aluminum, the energy cost can be reduced to less than 2000 kWh/ton.

Savings in transportation may be harder to attain, since they will require substantial changes in the American life-style. But the potential for energy conservation is considerable. When both direct and indirect energy costs are included, the automobile accounts for almost 21% of the total U.S. energy consumption! A Stanford Research Institute report indicates that automobiles directly accounted for 13% of all U.S. fuel consumption in 1968, with trucks using another 5%, airplanes 2%, trains 1%, and buses 0.2%. Reducing the weight of automobiles would result in large savings as shown in Figure 1-5.

The movement of people and materials on the nation's highways accounts for 76% of all the energy consumed in transportation. Airplanes account for 10% of the remainder, with even smaller amounts for rail (3.5%), shipping (4.8%), and fuel pipelines (5.2%). In the last two decades the fastest growing method of moving people and goods has been by airplane, which is also the most energy intensive as shown in Tables 1-2 and 1-3. The airlines' share of total intercity passenger traffic increased by a factor of 5 between 1950 and 1970, and their share of freight, sevenfold.

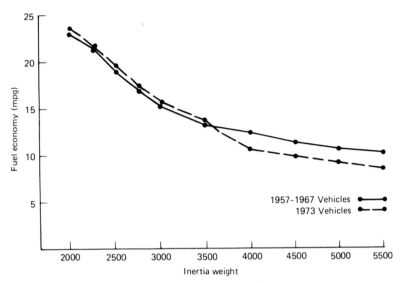

FIGURE 1-5 Disregarding minor efficiency fluctuations from model to model the fuel consumed per mile decreases remarkably as the car weight goes down. (Courtesy of Leo Schipper. Reproduced, with permission, from the *Annual Review of Energy*, Vol. 1. Copyright © 1976 by Annual Reviews Inc.)

From an energy standpoint Table 1-2 shows that it makes good sense to move toward bus and railroad mass transit. What makes mass transit so efficient is its large load factor, the average number of passengers carried per mile of actual travel. With increased passenger traffic, the mass transit efficiencies in Table 1-2 could be further improved by at least a factor of 2. The decline of both rail service and urban mass transit relative to automobiles and airplanes reflects the greater convenience and the savings in time that are usually possible, at a high-energy cost, with these two modes of transportation.

TABLE 1-2
Energy Intensiveness for Passenger Transport in Units of Joules per Passenger Kilometer

ITEM	URBAN	INTERCITY
Bicycle	0.339×10^6	
Walking	0.509×10^6	
Buses	6.28×10^6	2.71×10^6
Railroads		4.92×10^6
Automobiles	13.74×10^6	5.77×10^6
Airplanes		14.25×10^6

Source. E. Hirst, Oak Ridge National Laboratory.

TABLE 1-3
Energy Intensiveness for Freight Transport

ITEM	JOULES/Ton-km ($\times 10^6$)
Pipeline	0.763
Railroad	1.14
Waterway	1.15
Truck	6.44
Airplane	71.2

Source. E. Hirst, Oak Ridge National Laboratory.

D. QUESTIONS OF PHILOSOPHY

The behavior patterns of modern society have developed for a wide variety of political, historical, and economic reasons. There are a few common beliefs and philosophies that have strongly influenced these behavior patterns. For example, most of us are not yet truly aware of the long-term implications of the fact that the earth is a finite system. The most glaring example is that of oil. Figure 1-6 shows two estimates of the life cycle of the world's oil production. Q_∞ on this figure is the total available resource; the two quite different estimates of this quantity

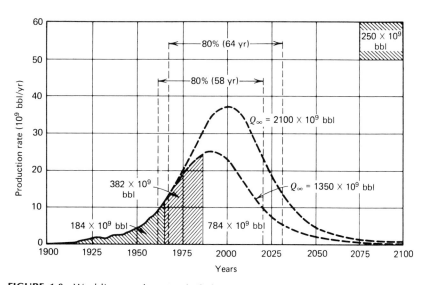

FIGURE 1-6 World's annual usage of oil plotted as a function of time. By the year 2000 the annual production of oil will have reached or passed the maximum. The "oil age" will only last a brief 100 years at most. The two different curves refer to different estimates of the total available world oil resource. [Originally published in the *Canadian Mining and Metallurgical Bulletin*, Vol. 66, No. 735, pp. 37–54 (1973).]

give a difference of only 10 years between the time of estimated peak production. This relationship between oil production and time was first pointed out over 25 years ago by the eminent geologist M. K. Hubbert. Little attention was paid at that time to his prediction that U.S. production would peak in the 1970 to 1975 period. The actual oil production rate has completely vindicated this prediction.

Many of the common minerals may soon be in the same state of affairs. It is mainly in this century that the mineral resources have been tapped to provide the base for our phenomenal industrial and economic expansion. The metal consumed in 30 years at present growth rates would be equal to that used in all previous history. Most of the world's prime mineral reserves have been found, and the United States and other countries have been mining them for years. By the end of this century the better grades of copper, tin, lead, mercury, and silver will have been seriously depleted. The mining of these better-grade ores has required little energy compared to the enormous drains that will be required to mine the low-grade, dispersed minerals in the future.

Shortages in other precious resources are imminent. Most everyone is now aware of the impending shortage of natural gas (see Chapter 3), but few are aware that our helium reserves will soon run out. We presently use it in such wonderfully wasteful ways as providing lift for balloons. Our reserves of this precious commodity will be mostly gone by the turn of the century. Helium is a national asset for cooling objects to low temperatures at which some metals become superconductors of electricity. In the future, superconducting, electric-power transmission lines cooled by liquid helium may be inexpensive enough to allow this technology to be implemented. Possibly, helium will be too expensive and too valuable to use outside of the research laboratory by the year 2000.

Little thought has been given to the needs of future generations. The world's supply of oil and natural gas was built up over a period of 600 million years by the conversion of solar energy through photochemical reactions. Society will use up 80 to 90% of this supply in 100 years leaving little left for the vital petrochemical industry. Our best fishing areas lie adjacent to the world's land masses. Continued pollution of these fishing areas with waste from the large population centers poses a real threat to the fishing industry. The excuse usually given for the rapid exploitation of our resources is that of practical economics. The time is near, however, when the nation's long-term needs and environmental considerations must be given as much weight in policy decisions as short-term, economic considerations.

The impressive accomplishments of modern science and technology have led most of us to believe that whenever a difficult problem such as the energy crisis arises, it will be overcome. It is certainly true that whenever the full power of our scientific and engineering talent is brought to bear on a problem the successes can be remarkable. The development of the atomic bomb during World War II and the Apollo flights to the moon are striking evidence of the ability of American science and technology to succeed with crash programs devoted to specific goals. On the other hand we tend to overlook the fact that the available supply of raw materials, talent, and resources was always far greater than was needed to accomplish these specific objectives. It is just as easy to overlook the vitally important, but slow, painstaking basic research that has gone into laying the foundation for these achievements. Consider the invention of the transistor, which came about only after a long series of developments, starting with basic solid-state theory in the 1930s and early practical solid-state devices in the late 1940s. Work on various types of associated devices continued in the early 1950s, particularly at the Bell Telephone Laboratory, finally culminating in the first transitor devices. The development came about as the natural consequence of a broad program of research, in many related areas and not as the result of a crash program designed to develop only a transistor. One cannot invent a transistor if its existence is not even known. Technological development must go hand in hand with a broad program of basic research.

An example of a problem that, thus far, has defied massive infusions of money and talent in the understandable hope for a quick solution, is cancer. Although much progress has been made, it has become quite apparent that this problem requires time-consuming fundamental re-search into the origins of cellular structure, the biochemistry of humans, the makeup of the body's immunological elements, and so on. Chances are that great advances will be made in the next 20 to 30 years, but this surely will not be easy.

Another extremely difficult problem is that of obtaining useful energy from fusion reactions. At one time this was looked on eagerly as the solution to all future energy needs. After years of research it is evident that attainment of practical fusion power will probably not come for a long time. Work on fusion should probably be regarded as a great scientific adventure. Present research is still a long way from obtaining the requisite temperature and density of the plasma (a gas of charged particles) that is needed. Before practical fusion reactors can enter the engineering stage, basic research must reach the point where considerably more energy comes out of the fusion device than is pumped in. Although those involved with fusion power generation are optimistic, and the reward of obtaining a practically infinite energy source provides great motivation, a long, difficult program of research lies ahead.

An even more difficult task is that of space travel to the stars. The nearest one is over 4 light-years away (1 light-year is the distance traveled by a beam of light, moving at 186,000 miles/sec, in one year, (that is, about 5.8×10^{12} miles) and is probably not sufficiently interesting to be worth visiting. The theory of relativity precludes travel at speeds equal to, or greater than, the speed of light. To get a rocket moving at speeds approaching that of light would require tremendous amounts of energy. The problems of maintaining living conditions during the trip, turning around to return, stopping upon returning, etc. seem impossible to solve at present. Furthermore, if we could somehow get near to the speed of light for most of the trip, the time span for the space travelers would be greatly reduced over that measured by some one on earth, so that when the space-trip travelers returned, every nontraveller would be much older. Present-day knowledge and technology is totally inadequate for this kind of a project.

Social considerations related to future energy sources are also important. Although the potential harm to society of energy shortages is great, the increasing utilization of technologically complex energy sources will be associated with a variety of difficulties. Dangers associated with fossil fuel development include the risks of spills from pipelines and offshore oil development, reclamation of land that has been strip-mined, and the environmental cost of shale oil development. Nuclear power implies a commitment to the management of waste products that long outlasts the power plants themselves. These kinds of social issues must influence decision making about energy use, especially since they favor more efficient energy use than would be dictated solely from economics.

This book is devoted to the subject of solar energy, our only important inexhaustible energy source. Its use offers the promise of an energy source that is clean, safe, and environmentally sound. The environmental advantages associated with solar energy applications are increasingly recognized as matters of paramount importance to society. It is now time to plan for a society that operates within a framework of environmental consciousness as well as social and economic stability. It is hoped that the time is past when the industrial world can think only of economic growth, with little concern for the needs of future inhabitants.

Solar energy is a renewable energy source in the sense that the sun will last forever as far as human life on earth is concerned. Its use also entails a minimal degradation of the environment. The use of solar energy adds no heat to the earth's biosphere; solar energy devices are called *invariant energy systems*. Some might add that the fuel is free. But, of course, collecting useful solar energy requires a considerable investment in capital equipment. Until recently, this initial cost for equipment and installation was so high that solar space heating could not compete with cheap fossil fuels. Another attractive feature for some is that for low-technology applications such as for space and water heating, the individual is in control of his or her own energy source.

This is of particular importance to those who wish to construct, install, and maintain their own systems.

Utilizing the sun as our energy source allows us a wide variety of options. These include direct uses such as the heating and cooling of homes, providing hot water and the generation of electricity through the intermediary of solar cells or solar-thermal approaches. These options form the primary subject matter of this book. In addition to the above there are such options as drying crops, producing food, growing kelp in the oceans as an oil substitute, and many other biomass applications. None of these options is new, although modern technological improvements have been needed to make most of them economically attractive.

Several features of solar energy must be considered in designing a particular application: (1) Its diffuse nature requires the use of relatively large surface areas for collection, (2) there are seasonal variations in the amount of energy available, (3) there is a daily variation of the available energy, which means that some means of storing energy is required, and (4) there are short-term fluctuations in the incident insolation, which will make the available solar power somewhat irregular at times.

Solar energy applications are not totally devoid of environmental impact. Conventional fuels must be utilized to extract, refine, and fabricate the basic materials needed for the construction of active and passive solar systems. This drawback can be overcome in the future as solar electric approaches are developed, providing the high-quality energy needed.

It is worth emphasizing that solar energy is technically feasible right now. Space heating and cooling systems are being installed at an increasing rate as the economic situation for solar systems improves each year. Perhaps the most impressive progress has been made in the area of solar hot water heating. With a combination of federal and state tax credits, the time required to pay back the initial system cost is usually between six to seven years. Almost as impressive has been the development of passive solar techniques for space heating in which the entire house serves as the collecting and storage system. These systems will provide for most of the building's heating needs over its entire lifetime with minimal maintenance and with no complicated apparatus to worry about. In addition to saving on fuel costs, passive solar homes are more comfortable to live in than an ordinary home, according to testimony from over 99% of passive home owners. This satisfaction appears to derive from the fact that solar houses are better insulated and are more bright and open than their conventional counterparts.

The case for solar electricity is not as well established, since developmental work is just beginning. Initial tests with pilot systems

FIGURE 1-7 Forest of mirrors. Faces the 61-m "power tower" at the Department of Energy's Solar Thermal Test Facility located at Albuquerque, New Mexico. This 5-MeW facility went into operation during the summer of 1978. (Courtesy of Sandia Laboratories, Albuquerque, New Mexico.)

indicate that a net energy gain will easily be obtained. One of the more promising possibilities is illustrated in Figure 1-7. Of course, much engineering development remains to be done before the real economic cost of solar electricity can be properly evaluated. Nonetheless, this is in sharp contrast to the situation with energy from nuclear fusion, which is far from being technically feasible at this point in time.

REFERENCES

1. Richard H. Montgomery and Jim Budnick, *The Solar Decision Book* (New York: Wiley, 1978).

   A simple, easy to read summary of the energy crisis is presented in Chapter 1 of this book. Contains a thoughtful discussion of the availability of energy in the future. A good presentation of the energy inflation rate is also given.

2. Wilson Clark, *Energy for Survival* (New York: Anchor Press/Doubleday, 1975).

   A detailed and clear presentation of our energy problems, environmental concerns, and alternative energy sources. The skillful, interesting prose makes this book both pleasurable and informative. The first three chapters are most relevant to the material discussed here.

3. C. Ruedisili and Morris W. Firebaugh, eds., *Perspectives on Energy*, 3rd ed. (New York: Oxford, 1982).

Contains an excellent series of essays written by experts. Probably the best single source of information on the topic of energy for the nonspecialist.

4. Jack M. Hollander and Melvin K. Simmons, eds., *Annual Review of Energy*, Vol. 1 (Palo Alto, Calif.: Annual Reviews, 1976).

   A selection of scholarly reviews of the literature in important areas of energy research. The interested reader should pay careful attention to the articles by Harrison Brown ("Energy in our Future") and Lee Schipper ("Raising the Productivity of Energy Utilization").

5. E. Cook, "The Flow of Energy in an Industrial Society," *Scientific American*, 134 (September 1971).

   This is one of many useful articles in the 1971 Energy issue of *Scientific American*. The reader may also find the article by C. M. Summers (p. 148) on the conversion of energy to be useful reading in this context.

6. Robert H. Romer, *Energy, An Introduction to Physics*, (San Francisco: Freeman, 1976).

   Chapter 1 of this text contains an excellent and more quantitative treatment of the energy crisis.

QUESTIONS

1. With 6% of the world's population, the United States presently has the following percentage of the world's energy consumption.

   (a) 50%   (c) 15%
   (b) 30%   (d) 6%

2. Suppose the ambient temperature outside your house is 7°C on a typical wintry day. Approximately what fraction of your heating needs can be saved by turning down the thermostats from 23°C to 18°C?

   (a) 1/6   (c) 2/3
   (b) 3/4   (d) 1/3

3. The least efficient mode of freight transport is the:

   (a) Railroad   (c) Truck
   (b) Airplane   (d) Waterway

4. The total amount of energy currently consumed in the United States per year is about:

   (a) 8×10^{16} J   (c) 80×10^{19} J
   (b) 8×10^{19} J   (d) 8×10^{12} J

5. The world's consumption of oil will peak near the year:

   (a) 1980   (c) 2030
   (b) 2000   (d) 2100

6. Suppose that 1000 solar homes are built in the United States this year and that construction grows exponentially, with a doubling time of two years. At what time in the future will the rate of construction be approximately equal to 1 million solar homes per year?

(a) 500 years
(b) 32 years
(c) 20 years
(d) 8 years

7. In the past 30 years the total energy consumption in the United States has been growing at a rate:

(a) Approximately equal to the population growth rate.
(b) Faster than the population growth rate.
(c) Slower than the population growth rate.
(d) No relationship.

8. The cost of electricity used in producing our common items such as machinery is about what percentage of the total production cost?

(a) 95%
(b) 50%
(c) 10%
(d) 2%

9. Which of the following is meant by the expression energy trap?

(a) That we are used to the availability of cheap energy.
(b) That we are used to the idea of energy being expensive.
(c) That much of our available energy resources are trapped in a way that renders them useless for practical development.
(d) That most of our energy resources are trapped in such a way that a very large amount of energy must be expended to release them.

10. As a primary source of energy, coal currently ranks:

(a) First
(b) Second
(c) Third
(d) Fourth

11. The utilization of a finite energy source such as oil follows the familiar bell-shaped curve (see figure). The area under this curve represents the total available resource. If the total area under the curve for our oil reserves is increased by 50%, what can you say about the corresponding change in the year at which peak production occurs?

(a) It will change by only a few years.
(b) It will change by at least 50 years.
(c) It will occur at a time that is 50% greater than that indicated on the figure above.
(d) It will not change at all.

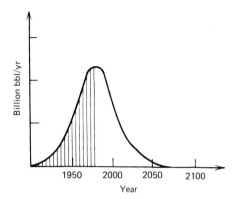

12. If the thickness of the insulation in the walls of a building is doubled, what can be said about the rate of energy loss through them (for the same temperature difference between the interior and the outside world)?

(a) It doubles.
(b) It remains the same.
(c) It decreases by a factor of 2.
(d) It decreases by a factor proportional to $1/\Delta T^4$, where ΔT is the temperature difference.

13. We are going to compare the electric bills during the month of December for two different all-electric houses. House number 1 operates with twenty 100-W lights, which are each kept on for an average of 10 hours per day. House number 2 has ten 100-W lights, which are each kept on for an average of 10 hours per day. Compare the electric bill for the two houses at the end of the month.

(a) The electric bill for house 1 is considerably greater than that for house 2.
(b) The electric bills for each house are about the same.
(c) The electric bill for house 1 is considerably less than that for house 2.
(d) The electric bill for house 1 is twice that for house 2.

14. Each of the following items concerns various approaches to energy conservation. Answer yes if the item describes an approach that will save energy, and answer no if the idea suggested will not help conserve our energy resources.

_____ (a) Lower building thermostats by 6°F.
_____ (b) Raise building thermostats by 6°F.
_____ (c) Decrease the amount of insulation in the walls.

_____ (d)   Increase the amount of insulation in the walls.
_____ (e)   Double the window area on the north side of buildings.
_____ (f)   Double the window area on the south side of buildings.
_____ (g)   Eat more carrots so that you can see in the dark.
_____ (h)   Manufacture lighter cars.
_____ (i)   Increase the number of pollution control devices on cars.
_____ (j)   Double the price of gasoline.

15.   In 1950, when the cost of oil was less than $3 per barrel, energy experts estimated that oil produced from oil shale would be profitable when the cost of conventional oil reached about $6 per barrel. In 1981 with the cost of conventional oil at about $38 per barrel, oil from oil shale was still not profitable. Explain why this is so.

16.   In the following problem you are to compare the item on the left with that on the right. Put an X in front of the *largest* one. An example is included to clarify the procedure.

_____ Your house	__X__ Empire State Building
_____ Energy expenditure per passenger mile for auto transportation	_____ Energy expenditure per passenger mile for air transportation
_____ Per capita energy usage in the United States	_____ Per capita energy usage in Western Europe
_____ The energy content of the U.S. proven oil reserves	_____ The energy content of the U.S. proven coal reserves
_____ The amount of radioactivity emitted routinely by a coal-fired electrical plant	_____ The amount of radioactivity emitted routinely by an average size nuclear reactor
_____ The insulation value of 6 in. of fiberglass	_____ The insulation value of 3 in. of fiberglass

17.   Suppose that 10% of the world supply of gold has been mined, in all of the earth's history up to now. If the doubling period for production of gold is 20 years, how many years will elapse before all of the gold has been mined?

18.   Some of the following are possible causes of the energy crisis and some are not. For each decide whether or not it is a possible cause and explain your line of reasoning.

(a)   Exponential growth in a finite system.
(b)   Energy may change form, but cannot be created or destroyed.

(c) Our reliance on finite, nonrenewable sources.
(d) Only a small percentage of the U.S. coal reserves are available for strip-mining.
(e) Our explicit reliance on the low cost of energy.
(f) The recent environmental protection laws and regulations.
(g) Our present inability to develop a working fusion reactor.
(h) The diffuse character of solar radiation implying the need for large-area collectors in order to obtain useful energy.

19. Next to each number write a letter giving the most reasonable matching answer. (Sometimes this may require an educated guess.)

_____ 1. Time for almost to-tal consumption of the world's oil re-serves
_____ 2. Proven U.S. oil re-serves
_____ 3. Time of formation of fossil fuels
_____ 4. Age of universe
_____ 5. Fossil fuels
_____ 6. Intensity of solar ra-diation

(a) Energy per square foot per hour
(b) 10 billion years
(c) 600 million years
(d) 1 million years
(e) Stored solar energy
(f) 10,000 years
(g) 200 years
(h) Proven reserves
(i) 30 million bbl
(j) 100 million bbl
(k) Energy per unit volume

20. Describe the relevance of each of the following with respect to the useful collection of solar energy.
(a) Renewable energy source
(b) An invariant energy source (with respect to the earth's biosphere)
(c) Need for storage of energy
(d) Large-area collectors

21. Devise a method whereby you can estimate the amount of energy used for space heating in an electrically heated house.(Hint. There may be some times of the year when no heating is required.)

22. The world is making increased usage of extra insulation in building design. Although this is good from an energy conservation point of view, it can lead to two important practical problems. What are they?

23. One possible method suggested for energy conservation in residences is to reduce the number of lights turned on. Discuss whether this course of action will result in a real energy savings during the winter season.

24. What advantages can you see in having to pay a deposit on bottles? Disadvantages?

25. If the thickness of insulation in the walls of a building is increased by a factor of 2, the heat loss through the walls will be reduced by almost a factor of 2. Suppose that the walls of a building are constructed of 4 cm of concrete. Discuss why 4 cm of concrete is not nearly as effective in reducing heat losses as 4 cm of fiberglass insulation. On the other hand, the concrete has the ability to store a large amount of heat as compared to the fiberglass. Discuss how this could be used to advantage in house design.

26. Glass windows on the south end of a residence are one good way to gain useful solar energy for space heating. However, care must be taken in using this approach. Tabulate and evaluate the possible disadvantages of this method of solar heating.

27. Doubling the amount of insulation in the attic of a house will effect a reduction in the upward heat loss of about one-half during the winter heating season. Will this extra insulation in the attic keep the house hotter or cooler in the summer? Explain.

28. One controversial subject at present is whether increased energy production is necessary to maintain a healthy economy. For example, one aspect concerns the number of jobs created in constructing a large electric generating plant versus the jobs created in constructing, installing, and maintaining the flat-plate solar collectors needed to produce the useful energy equivalent to that of the power plant. Research this topic and write a short paper. (*Suggestion.* In addition to the entries listed in the References, you will find some useful material in the 1977 issues of *Solar Age* magazine.)

29. The students at Henry Elementary School, St. Louis, Missouri, made the following suggestions to reduce our consumption of precious energy resources. Discuss the flaws (if any) in each suggestion.
    (a) Lower people's body temperature to 68°F.
    (b) Make it a rule that there must be at least two people in every big bed that uses an electric blanket.
    (c) Put more hot sauce in the food.
    (d) Don't have as many days of school.
    (e) Let birds fly around the house in summer to provide better air circulation.
    (f) Don't stay in more than one room at a time.
    (g) Eat more carrots so that you can see just as good with less light.

30. At present rates of consumption our fossil fuels will soon be depleted. How much faster than their rate of production are we consuming these nonrenewable energy sources?

(a) Twice as fast   (c) 1,000 times
(b) 10 times   (d) 1,000,000 times

2 energy

Energy comes in many different forms. In our everyday experience we see direct evidence of many of these different forms of energy. When we burn wood in a fireplace we are using the stored chemical energy of this fuel. The television waves captured by our TV set represent energy in the form of electromagnetic radiation. A rapidly moving automobile possesses an energy of motion that we call kinetic energy. The advent of the atomic bomb introduced us to the idea of nuclear energy, which is obtained by converting matter to energy. Other common energy terms with which we are familiar are work and heat. Precisely speaking, these last two are not forms of energy; instead they represent measures of the transfer of energy from one form to another. Many of our ideas about energy, such as our understanding of electromagnetic radiation and of the equivalence of mass and energy through the relation $E = MC^2$ (C is the velocity of light) are relatively new. Nevertheless, the concept of energy has proven itself to be both useful and durable.

In this chapter the concept of energy is examined in some detail. We begin by looking at several of the many common forms of energy. After a brief discussion of units, the energy concept is developed more fully through a number of practical examples. Finally the important subject of thermodynamics is discussed. Most solar energy applications utilize some type of thermal process in order to obtain useful heat, mechanical work, or electrical power. To understand how this is best accomplished it is necessary to gain an appreciation of the limitations set by nature through the first and second laws of thermodynamics.

A.   WHAT IS ENERGY?

Although energy is a very important concept, it is difficult to state a precise definition. Energy is not an object that one can see or feel, but a concept that helps describe the state of an object or a system. For our purposes a useful definition of energy is that it is the capacity to do work. (Later on in the chapter this statement will be qualified; appropriate conversion devices must be available, and the temperature of the surroundings must be sufficiently low.) For example, by the use of properly designed engines, the stored chemical energy in gasoline can be converted into useful mechanical work. It is up to human beings to find ways of converting the potential capability of our energy sources into useful work for society in an efficient, economical, and environmentally clean manner. For most energy sources these last two requirements are not always compatible.

A better understanding of energy can be obtained by taking a brief look at some of the many forms in which it exists. Figure 2-1 shows a stream of water cascading over a waterfall. The water at the top of the

FIGURE 2-1 A stream of water flowing over a waterfall illustrates the concepts of potential and kinetic energy.

waterfall has a large amount of stored, or potential energy. This potential energy was gained because the sun's rays, striking the oceans, caused the water to evaporate and rise in the air. Some of this potential energy was lost as the water fell to the earth in the form of rain. The water at the top of the waterfall still has a large potential energy because of its high position relative to the oceans. As the water spills over the waterfall it picks up speed. In this case the force of gravity is doing work on the water molecules. The work done by the gravitational force is converted into energy of motion of the water. This energy of motion is termed kinetic energy. In physics terms, the change in kinetic energy of the water molecules in going from the top of the waterfall to the bottom is exactly equal to the work done by the gravitational force. Under control, as in the case of a dam erected to generate hydroelectric power, the fast moving water at the bottom of the dam can be utilized to turn the blades of a turbine. The latter can then be used to generate electricity.

Many forms of matter contain large amounts of stored energy based on their chemical compositon. The oil, shown gushing from the well in Figure 2-2, represents one of nature's great stores of chemical energy. Coal, natural gas, and tar sands are other important sources of this form of energy. In Chapter 1 we saw that fossil fuels presently represent the world's largest source of energy.

The strongest force in nature is the nuclear force which is responsible for holding the individual particles of a nucleus together. The

FIGURE 2-2 Oil is a prime example of stored chemical energy. It took nature hundreds of millions of years to make this finished product.

energy associated with the nuclear forces between particles is extremely large. In a nuclear reactor some of this nuclear potential energy is released and converted into kinetic energy. This resulting kinetic energy can then be converted to useful electric power. This electrical energy can be stored by means of a battery. In this case an electric charge is maintained at a given voltage (electric potential).

Finally, let us consider the concept of thermal energy and heat. These are topics with which we will be much concerned in the rest of the book. Thermal energy is the energy an object possesses because of the motion of its molecules. The larger the thermal energy, the greater the kinetic energy of the molecules of the substance. What we call heat is the transfer of thermal energy from one place to another. Hot objects also give off radiative energy in the form of electromagnetic radiation. This topic is discussed separately in the chapter on solar radiation.

Why is the concept of energy so important? The answer is related to the ability of energy to change from one form to another. For example, the water at the top of a waterfall has a certain amount of (gravitational) potential energy. As the water flows over the waterfall and rushes toward the bottom, this potential energy is transformed into kinetic energy. This kinetic energy can be used to run a turbine and generate electricity. In turn, this electrical energy can be converted into heat, light, sound, and so on.

But all of this would be considerably less useful if it were not for the fundamental law of physics known as the law of conservation of energy, which states, "Energy cannot be created or destroyed, but can only be transformed from one form to another." This means that while the amount of any particular form of energy may change, the total amount of all forms of energy in any isolated system must remain a constant. Without this law of nature, the disappearance of a quantity of energy in one place could not be related to the appearance of a similar amount of energy somewhere else. If it were not for the law of conservation of energy, two different manifestations of energy would involve two different concepts.

In the first chapter conservation of energy was discussed as a possible strategy to help soften the deleterious effects of the growing energy crisis. In that context, conservation implied a reduction in the consumption of our energy resources. Now the word conservation is used in a different sense. If a quantity is conserved, this means that the value or amount of this quantity does not change no matter what physical processes take place. Used in this way conservation of energy implies that while the energy content of an isolated system may change its form, the total amount measured in some appropriate unit remains constant.

In what follows we give a few simple examples of energy calculations. These are given in order to provide the reader with an intuitive feel for energy and the relative size of various units of energy. To do this it is first necessary to digress slightly and discuss the topic of systems of units.

B. SYSTEMS OF UNITS

A quantitative discussion of energy requires the specification of a system of units. Intuitively we recognize that units should be specified for the three fundamental dimensions of length, mass (roughly speaking, the weight of an object), and time. Once these three are specified all other mechanical quantities can be expressed in terms of some combination of length, mass, and time. As other physical concepts such as electricity are introduced, one might suppose that fundamental dimensions (such as electrical charge) might be needed. Instead, it turns out the above three dimensions are sufficient to quantitatively describe all known physical quantities. It is also common practice to use a fourth quantity, temperature, as another basic dimension rather than defining it in terms of mass, length, and time.

In terms of these fundamental dimensions there are three common systems of units. These are the MKS system, in which the meter (m), kilogram (kg), and second (sec) are the basic units; the CGS system, in which the centimeter (cm), gram (g), and second are the units; and the English system, comprised of the foot (ft), slug [or pound (lb)], and second. The choice of which system to use depends on the application. The MKS system is advantageous when considering electrical quantities that are expressed in volts (V) and amperes (A). For atomic and nuclear phenomena the CGS system is commonly used. Because of historical precedent, most common quantities are still expressed in English units, although this is now changing slowly as the United States moves toward complete use of the metric system. Metric system units will be adopted as the standard in this text, but English units will often be included for clarity.

A proper definition of mass requires a study of dynamics, which is beyond the scope of this book. However, an operational definition can be made in terms of a standard block of material. The international standard is a cylinder of platinum-iridium that is defined as 1 kilogram (kg) or 1000 grams (g). A beam balance or similar devices can then be used to compare the standard mass with any other object. Secondary standards of greater and smaller mass can be constructed by comparison with the standard mass. In the English system the weight of an object is expressed in pounds. One kilogram is equivalent to the weight of about 2.2 lb of material. The mass of an object differs in a fundamental way from length and time in that the amount of material in an object comes in discrete amounts. All matter is made up of the fundamental constituents of an atom—electrons, protons, and neutrons. Length and time, on the other hand, are infinitely divisible or continuous.

The metric system defines length in terms of a standard meter. This was originally defined as the distance between two parallel scribe marks on a bar of platinum-iridium maintained at a temperature of 0°C. This standard meter is housed in the International Bureau of Weights and Measures, near Paris, and secondary standards have been distributed to national standards agencies throughout the world. As a way of improving on the accuracy of the standard meter it was agreed in 1961 to define it in terms of a natural unit based on atomic radiation. The meter is now defined as 1,650,763.73 wavelengths of the orange light from krypton gas, a definition that has the tremendous advantage of being precisely reproducible everywhere. This atomic standard is 100 times as precise as that obtained from measuring the distance between scribe marks on a meter bar. In English units the meter is about 39.37 in.

Webster's Dictionary defines time as the period during which an action or process continues. Thus, the concept of time is related to the idea of periodic repetition. The essential requirement for a time measurement is to use a process that repeats in a regular, precise pattern. One choice for a standard of time is the mean solar day, the average time interval between successive passages of the sun through a given meridian. The second is then defined as 1/86,400 of the mean solar day. To improve the accuracy of time measurements, recourse was made in 1967 to an atomic standard that now permits an accuracy of one part in 10^{12} to be obtained.

Temperature is a measure of the average kinetic energy of all the molecules of a substance. There are a large number of ways to define the concept of temperature. A very imprecise method is to utilize our senses to define what is meant by the temperature of an object. This approach discriminates between the temperature of boiling water, ice, and a hot flame, but does not provide a useful quantitative measurement. For more precision it is possible to take advantage of a number of properties of matter that vary with temperature, and these can be used to construct thermometers. For example, mercury expands more than glass when heated, so that the length of the liquid column in a mercury-glass thermometer is a measure of the temperature. Another example consists of two strips of different metals that are joined together side-by-side. When a temperature change occurs, the pair of strips bends owing to the different rates of expansion in the two metals. The higher the temperature, the greater the deflection. When cooled, the bimetallic strip bends in the opposite direction.

To use a thermometer it is necessary to establish a scale of temperature. It is customary to utilize the boiling and freezing points of water to define the scale of temperature. If we call these two points 100° and 0°, respectively, we have the Celsius scale of temperature. If we call these two points 212° and 32°, respectively, we have the Fahrenheit scale of temperature. On the Fahrenheit scale there is a 180° difference between the freezing and boiling points of water, whereas the difference on the Celsius scale is 100°, or to put it another way, one Celsius degree is 9/5 as large as a Fahrenheit degree. Thus,

$$T_F = 32 + \frac{9}{5} T_C \qquad (2\text{-}1)$$

and, rewritten for T_C,

$$T_C = \frac{5}{9} (T_F - 32) \qquad (2\text{-}2)$$

where T_F is the temperature in Fahrenheit degrees and T_C the corresponding temperature in degrees Celsius.

In many applications it is necessary to use a different temperature scale, where zero degrees corresponds to no molecular motion at all. This is necessary in order to simply express the amount of radiation emitted by a hot, glowing object, for example. The new temperature scale is called the Kelvin scale of temperature. In this case, absolute zero on the Kelvin scale corresponds to $-273°C$ as indicated in Figure 2-3. Theoretically, at zero Kelvin, all molecular motion stops and a body contains no thermal energy. Any temperature on the Celsius scale can be converted to the Kelvin scale simply by adding 273. On the

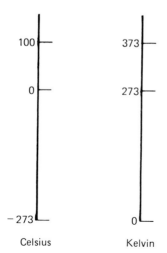

Celsius   Kelvin

FIGURE 2-3 Comparison of the Celsius and the absolute (Kelvin) scales of temperature.

Celsius scale the boiling point of water is 100°C, and this corresponds to 373 Kelvins. (In using the Kelvin scale it is customary to just say Kelvins, not degrees Kelvin.)

C.   ENERGY EXAMPLES

The rate at which energy is converted from one form to another is often as important as the total energy converted. The term *power* is technically reserved for this rate. Thus power, P, can be expressed as

$$P = \frac{\text{work}}{\text{time}} = \frac{W}{t} \tag{2-3}$$

In the MKS system where work is expressed in joules, the unit of power is the joule per second, which is called a watt (W). The watt is rather small for most industrial purposes; even an electric clock requires several watts of power. A more practical unit is the kilowatt (kW), which equals 1000 watts. Large power plants produce useful electrical energy at rates of the order of 1000 million watts, or 1 million kilowatts, or 1000 megawatts (1000 MW).

EXAMPLE 1

Electric bills are commonly defined in terms of another practical unit, the kilowatt hour (kWh). Let us express a kilowatt hour in terms of joules.

$$1 \text{ kWh} = 10^3 \text{ Wh} \times 3600 \text{ sec/hr} = 3.6 \times 10^6 \text{ watt-sec}$$

Since 1 W equals a joule/second, this gives

$$1 \text{ kWh} = 3.6 \times 10^6 \text{ J}$$

It would be most logical to use the joule as the energy unit for the many applications involving the transformation of energy from one form to another. Unfortunately, things are not always done in the most logical manner. Before the middle of the nineteenth century the equivalence of the different forms of energy to heat was not understood. For a long time heat was thought to be a fluid, called caloric, which could flow from one object to another. This resulted in different units for heat than those used to describe mechanical work. The modern view of heat is described in more detail later in this chapter. For the present it suffices to say that heat is the energy of motion associated with random motion of the molecules of a substance.

Since heat was originally thought of as a special fluid, it was given its own unit. In the metric system this unit was called the calorie (cal) and was defined as the energy required to raise the temperature of 1 gram of water 1°C. In the English system of units, heat was historically measured in British thermal units (Btu). This unit is defined as the energy required to raise the temperature of 1 lb of water 1°F. Formally, the unit of heat can be related to the change in temperature of a substance by the following equation:

Quantity of heat = Specific heat × Mass × Change in temperature,

$$Q = CM\Delta T \tag{2-4}$$

The constant C is called the specific heat capacity and is a function of the type of material and the system of units. The heat capacities of a number of common materials are given in Table 2-1.

Since heat is only one form of energy, the heat units of British thermal units and calories can be related to work units such as joules. It was the pioneering work of Count Rumford and Joule that first confirmed the quantitative relation between the two different energy units. Careful experimental measurements established that 1 cal was equal to about 4.18 J. The calorie has now been redefined to be exactly 4.184 J. Thus, the specific heat of water is 4.184 kJ/kg-°C. Energy will usually be expressed in joules throughout this text. However, since it is still common practice in the United States to express energy in British thermal units, it is useful to be able to convert readily between the two units. To do this note that 1 Btu is equivalent to 1055 J. For many comparisons it suffices to multiply the number of British thermal units by 1000 to get the equivalent number of joules.

EXAMPLE 2

The temperature of 454 g of water is raised from 20 to 40°C. How much heat is involved in the process? In metric units,

$$Q = 4.184 \text{ kJ/kg-°C} \times 0.454 \text{ kg} \times (40\text{-}20)\text{°C} = 37.99 \text{ kJ. } (37{,}990 \text{ J}).$$

TABLE 2-1
Heat Capacities of a Number of Common Materials

SUBSTANCE	CAL/g-°C	kJ/kg-°C
Water	1.0	4.184
Ice	0.5	2.09
Brick	0.2	0.84
Cement	0.16	0.67
Rock	0.2	0.84
Plywood	0.29	1.21
Softwoods	0.33	1.38
Glass	0.24	1.0
Air	0.17	0.71
Aluminum	0.22	0.92
Iron	0.11	0.46

To calculate the heat flow in English units we note that 454 g equals one pound. Also, from Eq. 2-1,

$$\Delta T(°F) = \frac{9}{5} \times T(°C) = \frac{9}{5} \times 20 = 36°F$$

$$Q = 1 \text{ Btu/lb-°F} \times 1 \text{ lb} \times 36°F = 36 \text{ Btu}$$

The amount of energy used for space and water heating is an important item in estimating energy needs. Let us first look at the amount of energy in a gallon of oil and calculate the cost of heating a home in this manner.

EXAMPLE 3

Make a rough estimate of the cost of heating with oil. Assume that the cost of oil (delivered to you) is $60 per barrel (bbl). To do this exercise we need to know that there are 5.6×10^6 Btu in a barrel of oil. Also, one barrel of oil is equivalent to 42 gallons. The cost per gallon is then

$$\$60/42 = \$1.43 \text{ per gallon}$$

To find the cost per unit of energy note that 1 kWh = 3.6×10^6 J = 3.6×10^6 J/1055 J/Btu = 3412 Btu. Thus

$$\frac{\text{cost}}{\text{kWh}} = \frac{\$60.00/\text{bbl}}{5.6 \times 10^6 \text{ Btu/bbl}} \times 3412 \frac{\text{Btu}}{\text{kWh}} = 0.037/\text{kWh}$$

However, oil furnaces are not 100% efficient. A typical furnace efficiency might be about 70%, so that the above cost estimate must be divided by 0.70.

$$\left[\frac{\text{cost}}{\text{kWh}}\right]_{\text{heating}} = \left[\frac{\$0.037}{0.70}\right] = \$0.052/\text{kWh}$$

Of particular importance to us is the amount of energy contained in the solar radiation coming from the sun. This is usually expressed in terms of the solar energy incident on a unit area. Commonly this will be in joules per square centimeter (J/cm^2) or in kilojoules per square meter (kJ/m^2). It is instructive to calculate the amount of solar energy that annually strikes the roof of an average house in the United States.

EXAMPLE 4

The average annual solar radiation on a horizontal surface varies considerably across the country. A value of 5×10^9 J/m^2 is a reasonable estimate. Assume a roof area of 1500 ft^2, which in metric units is

$$1500 \text{ ft}^2 \times \frac{1}{10.76 \text{ ft}^2/\text{m}^2} = 139.4 \text{ m}^2$$

$$\begin{bmatrix} \text{Solar energy} \\ \text{striking house} \end{bmatrix} = 5 \times 10^9 \frac{\text{J}}{\text{m}^2} \times 139.4 \text{ m}^2 = 697 \times 10^9 \text{ J}$$

The typical annual space heating requirement for a house this size may vary from 50 to 100 $\times$ 10^9 J, depending on the climatic region. Thus, the solar radiation striking the roof of a house is 7 to 14 times the heating need. Of course, we can not usually collect energy during the summer for winter heating so that the above number should be divided by a factor of roughly 3 to correct for this. In addition, the generally recommended approach to solar heating nowadays is to use some type of passive approach. If we have 300 ft^2 of south-facing glazing, then the above number is reduced to

$$\frac{697 \times 10^9 \text{ J}}{3 \times 5} \cong 46 \times 10^9 \text{ J}$$

where the factor of 5 is due to the smaller area of glazing (compared to 1500 ft^2 in the above calculation). The actual efficiency of collection of solar energy and conversion to useful heat for the building will be somewhat less than 100%. The end result is that only about two-thirds of the building's heating needs will customarily be provided for through the solar augmentation.

How does the United States use its energy? The output uses of energy are shown schematically in Figure 2-4. Industrial uses and transportation account for almost 72% of the consumption of energy resources.

It is also instructive to combine the various output uses into specific tasks. Looking at the end use of energy in this way confirms the concern about the automobile, which accounts for nearly 13% of the country's consumption of energy resources. Space and water heating account for almost 25% of the total energy use. The application of solar energy techniques to space and water heating can result in a considerable savings of nonrenewable energy resources. Another area where solar energy could have a large impact involves industrial use of process steam; this category accounts for nearly 17% of the nation's total energy use.

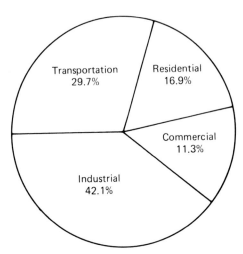

FIGURE 2-4   Graphic representation of the end use of energy in the United States.

D.   ENERGY COST OF PRODUCING ENERGY

A central problem that must be recognized and overcome is that it takes a certain investment of our energy resources to convert the energy contained in other resources into a useful form. For example, to obtain oil from shale rock, the machines that do the needed work of separation consume oil (or the energy equivalent from some other source such as coal). To be profitable, the amount of oil consumed to extract the oil from the shale must be considerably less than that obtained by the separation process.

The failure to recognize the above fact can have very severe consequences. In the mid nineteenth century a British company took over the luxury liner Great Eastern, and decided to make use of it for carrying coal from Australia to England. This huge ship, weighing 19,000 tons, and equipped with bunkers capable of holding 12,000 tons of coal, was designed to travel to Australia and back without refueling. Unfortunately, the designers were extremely remiss in their estimates of the coal needs of the ship's engines. It was found that for the Great Eastern to make a round trip to Australia, the ship would need three times more coal than it could carry. The company folded and the ship was broken up for scrap metal in 1888.

The above example illustrates an important lesson. The Great Eastern required more energy to operate than it could carry. The same constraint must be imposed on the future utilization of our energy resources. We have taken considerable advantage in the past of energy sources that were relatively easy to obtain and convert to practical use.

However, as more complex energy technologies are implemented, care must be taken to ensure that the energy resources consumed are not as large as those obtained. Until recently it appears that this aspect was not seriously considered in the evaluation of potential energy reserves. To better illustrate this point, let us look at a few practical examples.

In 1930 the oil industry found 275 bbl of oil for each foot of exploratory drilling, with the average find at a depth of 3000 ft. Today this figure has fallen to 35 bbl with the average well depth at 6000 ft. Digging exploratory wells to greater depths requires the use of more energy. If we add in the energy "costs" in the future due to the problems of offshore drilling and of constructing long pipelines, it becomes apparent that the net gain per barrel of oil produced will be significantly reduced. Despite these increased technological difficulties, the net gain from each unit of oil produced will probably always be a large fraction of the total amount of energy available from the oil obtained in this way.

The same statement cannot be made with such certainty about our other fossil fuels. Strip-mined coal is the easiest to obtain—particularly if proper reclamation procedures are not followed. However, the largest fraction of our coal reserves lies in deeper veins and will require a significant energy input to mine. There is even greater uncertainty concerning the process of extracting oil from the large potential reserves of shale rock located in the western part of the United States. Oil shale is a sedimentary rock that contains kerogen, a solid, tarlike organic material. This shale rock requires extensive mining, heating, and processing in order to release the oil from the kerogen. It is not certain yet that this extensive working of the shale rock will actually yield a net return in energy and, even if it does, the fractional return will probably be low. As another example, the Athabasca tar sands in Central Canada are estimated to contain about 500 billion bbl of oil. However, it requires the consumption of about 90 bbl of oil in order to obtain 100 bbl of oil!

Solar energy applications are in a more favorable position. A typical flat-plate solar collector will gather enough useful net energy in one to two years to cover the energy resources consumed in the manufacturing process. The situation for most of the large-scale approaches to solar electricity also appears to be equally promising.

A detailed assessment of the amount of energy resources consumed in the production process must also take into account the concept of efficiency. No real device, such as an automobile engine, can extract all the energy from a potential energy source and convert it into useful work. Not only are there operating losses such as those due to friction but even a perfect heat engine has a useful output that is limited by the second law of thermodynamics. The rationale and implications of this law of nature are discussed in the next section. For our present purposes notice that the useful work output from a heat engine such as the gasoline engine in cars or the steam engine in a boat or train is limited by the second law of thermodynamics to be no greater than 30 to 50% of the available energy potential. The actual output will often be considerably less than this. The efficiency η of a device is defined as

$$\eta = \frac{\text{useful work output}}{\text{energy in}} = \frac{W}{Q_{in}} \qquad (2\text{-}5)$$

Thus, the useful work output equals ηQ_{in}. For example, if an oil-burning engine has an overall efficiency of 15%, then the net work output from 100 bbl of oil would only be the total energy equivalent of 15 bbl. In the earlier example of the Athabasca tar sands, it was assumed that the efficiency in using the 90 bbl of oil to obtain the 100 bbl is the same as the efficiency of utilization of the produced oil.

What about the fraction of energy $(1 - \eta)Q_{in}$ that is not converted into useful work? For example, a nuclear-powered electrical generating plant typically has η at about 0.30 (30%). The other 70% of the thermal energy of the reactor is called waste heat and is expelled into the atmosphere, a nearby river, or the ocean. The resulting thermal pollution can have serious consequences for the environment. This waste heat does not have to be thrown away in this manner. It could be transported away and used for useful heating purposes, as is the case in Sweden where district heating is prevalent. As the cost of energy rises in the future, energy conservation in this way will undoubtedly become more popular.

The energy cost of producing electricity with nuclear reactors was the subject of considerable debate for several years. A large input of energy is required for the mining and milling of the natural uranium, to enrich the amount of ^{235}U isotope (see Chapter 3), to about 4% from its naturally occurring value of 0.7% and to construct and operate the reactor. Several analyses of these steps now agree that the electrical energy required is approximately 25% of that obtained from the burning of ^{235}U in an American light-water reactor.

E. THERMODYNAMICS

The generation of solar electricity is now an established fact. Chapters 10 and 11 are devoted to this topic. Except for solar cells, all other realistic solar-electric approaches involve the conversion of heat energy to electricity. This conversion process requires an engine working in a cyclic fashion. To do this some working material, say hot water or steam, must undergo a series of changes of pressure, volume, and temperature. When this happens we say that a thermodynamic process has taken place. The entire subject is complicated because of the fact that we are not dealing with a simple system of a few molecules, where the laws of mechanics can be applied. Instead, we are dealing with incredibly large systems of particles so that only the average behavior of such a system can be described. Despite the complexity, the description of such systems was fully understood by the end of the nineteenth century. Practical thermodynamics, from an engineering standpoint, can be quite complex and detailed. For our purpose, which is to understand the basic requirements for generating electricity, comprehension of a few fundamental concepts will be sufficient.

It was pointed out earlier that the usefulness of the concept of energy is due to the fact that the total energy of a system is conserved. When energy is used to perform useful work, what occurs is that energy is converted from one form to another. When gasoline is burned in a car, the stored potential energy (chemical) in the fuel is converted into useful work—the motional energy of the auto. But, this example raises a serious question. A quantitative measurement of the useful work done on the car by burning the gasoline will show that only a small fraction of the stored energy in the gasoline has been converted to useful purposes. The remainder has been expended in friction or has been exhausted into the atmosphere. This seems like a shameful waste. Can human beings do something about this wasteful use of energy?

In this connection it is useful to review the efficiency of various ways of generating electricity. Remember that the efficiency of a device is defined as the ratio of useful work output to the total energy put into it. The efficiency of a typical coal-fired, electric power generating plant is typically about 40%. This means that 60% of the available, stored energy in the coal has been rejected to the outside world. Nuclear reactors commonly have efficiencies in the 32% range, giving an even larger fraction of energy rejected. On the other hand, when water passes over a dam and is used to generate electricity, it is found that almost 90% of the gravitational potential energy of the water is converted to electrical energy. This is a tremendous improvement over the other ways of generating electrical power and leads us to ask exactly what is responsible for such wide variations in the efficiency of power generation.

One possible explanation might be that there has been a breakdown in the law of conservation of energy; however, since there is no evidence for this in any other measurement in physics, we rule out this possibility. Furthermore, we also rule out friction losses in the turbines, and so on. Good design can often reduce the friction losses to a few percent, so that this cannot be the reason for the relatively low efficiency of coal and nuclear power plants as compared to hydroelectric power. Instead, the solution to this puzzle lies at a more fundamental level. The low efficiencies are due to a fundamental thermodynamic limit on the efficiency of conversion of energy to useful work by any process involving a transfer of heat. This limit constrains the possible efficiency of conversion of heat to work, such as in the generation of electricity. However, there is no such constraint on the reverse process, since work can be converted to heat with an efficiency of 100%.

The detailed explanation of this process involves some very subtle ideas. Before beginning, it is useful to first discuss the concept of conservation of energy carefully in the context of processes involving heat. As noted earlier, until the middle of the nineteenth century, it was thought that heat was somehow proportional to the amount of a fluid (called caloric) that was contained in a substance. This caloric theory of heat was able to explain many phenomena. For example, if more caloric flowed into a substance, then it should become warmer, exactly as observed.

However, there were problems. If heat is a fluid and is squeezed out of a piece of material, there should be less heat left in the material. This results in the material having less capacity to hold heat. Yet no such effect was ever found. The first scientist to quantitatively show that heat was just another form of energy was Benjamin Thompson (Count Rumford). He became interested in the question of what exactly is heat while supervising the boring of cannons. This was performed by a number of oxen, who traversed a circular motion about the cannon in order to turn the mechanical device responsible for boring the cannon. By careful computation (for these circumstances) he found that the heat generated in the cannon, which was obtained from the rise of tempera-ture, was proportional to the mechanical work done by the oxen in a fixed time period—no matter how many hours or days the process had been going on. He concluded that heat is associated with microscopic motion. According to the caloric theory, the chips bored from the brass cannon barrel would be hotter than the barrel because the bore would squeeze the caloric out of the barrel along with the brass chips; in fact, Count Rumford discovered that both the bored barrel and the chips were at the same temperature. Several years after the experiments of Count Rumford, Joule made a careful series of measurements in order to determine quantitatively the relation between heat energy in calories and mechanical energy in joules. The present-day accepted equiva-lence is 1 cal = 4.184 J.

E.1 The First Law

It is now understood that heat is a form of energy. More precisely, it is the transfer of energy by molecular collisions. Scientists call the energy content of a substance its internal energy and save the word heat for the gain or loss of internal energy brought about by molecular collisions. For example, if a hot iron poker is dipped into cold water (Figure 2-5), the particles of iron lose energy while the water molecules gain energy. Energy has been transferred from the hot poker to the water molecules

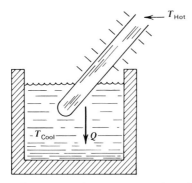

FIGURE 2-5 The energy transferred by molecular collisions between the hot poker and the cooler water is called heat. The energy content (due to random molecular motion) of the water or the hot poker is termed the internal energy of the substance. The flow of heat, Q, is from T_{HOT} to T_{COOL}. At equilibrium the temperature of the water and the poker will be identical. This resultant equilibrium temperature will be intermediate between the two initial temperatures.

as a result of a large number of molecular collisions. The internal energy of a substance is proportional to the energy of the random motion of the molecules of the material. In turn, this motional energy is proportional to the parameter we have called temperature. In other words, the energy content of a substance is proportional to its absolute temperature (or vice versa).

The statement that energy is conserved is an empirical law of physics. When this law is extended to include heat and internal energy, it is called the first law of thermodynamics. We demonstrate the use of the first law with the aid of the simple system shown in Figure 2-6. Suppose that a quantity of heat is added to this system. For example, this could be done by using the flame of a bunsen burner. This additional heat will warm up the gas causing its molecules to move faster. In other words, the internal energy of the gas will have increased as measured by its increased temperature. At the same time, the increased thermal motion of the gas molecules causes the pressure on the piston to increase. If the piston is held fixed, no work will be done and all of the heat energy ΔQ will go into increasing the internal energy of the gas, resulting in a temperature increase from its initial value to a final value T_1. On the other hand, if the piston is free to move, a certain amount of work ΔW will be done in displacing it upward against the force of gravity. The heat energy ΔQ will then be divided in some

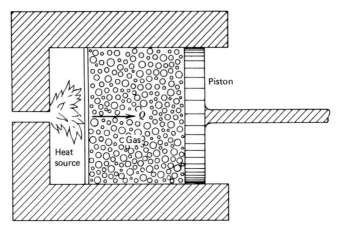

FIGURE 2-6 Apparatus demonstrating the first law of thermodynamics. If the piston is held fixed no work is done, and Q equals the change in internal energy of the gas. However, if the piston is free to move, then some work is done and conservation of energy tells us that the resultant increase in internal energy of the gas is smaller than for the case where the piston was fixed.

manner between the work ΔW and the change in internal energy. In this case, the first law, which expresses conservation of energy, tells us that the increase in internal energy will be smaller than in the first case, so that the final temperature of the gas, T_2, will be smaller than T_1.

E.2   The Second Law

Let us now return to the question about the efficiency of generating stations that was raised at the beginning of this section on thermodynamics. Consider a specific example of one type of heat engine used for large-scale electric power generation, the steam turbine. The basic idea can be understood with the aid of Figure 2-7. Steam is heated to a condition of high temperature and pressure and allowed to expand against the blades of the turbine. This expansion causes the blades to rotate, so that heat energy has been converted to useful work, in a manner analogous to the linear motion of the piston (Figure 2-6), under the influence of the expanding gases. The source of heat could come from nuclear reactions, the burning of oil or coal, hot water or steam from a geothermal source, or from some type of solar energy collector. In the case of the turbine not all the energy contained in the hot steam is converted to useful work (the rotating blades). The outlet gas that is at a finite temperature has a quite significant amount of internal energy. Consequently, typically only 30 to 40% of the incoming energy is actually converted to useful work. This performance is to be compared with the efficiency of the turbine of a hydroelectric plant such as that shown in Figure 2-8, which is almost 90%.

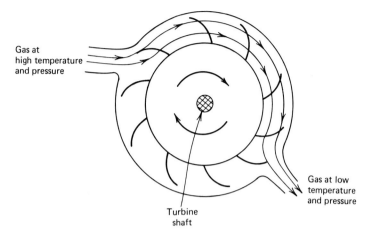

Gas at
high temperature
and pressure

Gas at low
temperature
and pressure

Turbine
shaft

FIGURE 2-7 Example of a steam turbine used in thermal power plants to produce electricity. The hot gas expands against the turbine blades to produce rotation, which in turn leads to electric power. A typical efficiency for a coal-fired power plant is 40%.

Why is the turbine efficiency so great in the hydroelectric case? The answer is found in the fact that the motion of the falling water is ordered, with the water molecules having a dominant velocity component along the stream flow, so that all of the water falls in a given direction. This is not true for the internal energy of molecules in a hot

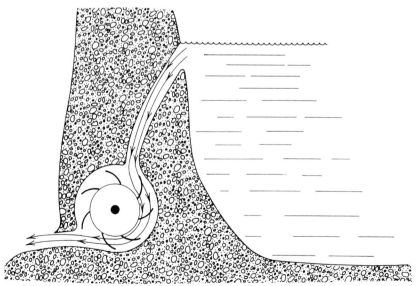

FIGURE 2-8 Turbine in a hydroelectric power plant. The overall efficiency for converting the stored potential energy of the water to electric power is of the order of 90%.

gas where the motion is random, or disordered. The results of many investigations of thermodynamic processes have shown that nature has a general rule stating that there is a tendency toward disorder, provided that one deals with large enough samples of matter where more than a few molecules are involved. Thus, the rule is

$$Order \rightarrow Disorder$$

Because no exceptions to this tendency have been found (for a closed complete system), it is accepted as a law of nature and is called the second law of thermodynamics.

Since mechanical energy of motion is more ordered than is thermal energy, it can always be completely converted into other forms, resulting in the very high efficiency found in the hydroelectric turbine.

On the other hand, disordered thermal motion cannot be completely converted into ordered mechanical motion. In the case of the conversion of thermal energy in the steam turbine of Figure 2-7, only a fraction can go to ordered mechanical energy while the remainder must be in the form of even more disordered thermal motion of the ejected exhaust gases. In other words, only a portion of the input energy in a thermal electric generating plant can be converted to useful work (in the form of electric power), and the efficiency of such plants is much lower than for hydroelectric generation.

The above description of the second law of thermodynamics in terms of ordered and disordered motion is rather abstract. Thanks to the work of a brilliant young French field artillery officer and physicist, Sadi Carnot, the second law can be restated in an entirely different manner. To do this it is necessary to introduce the concept of a thermodynamic cycle. For example, the fluid used in a steam turbine undergoes a cyclic process in which the steam starts out in some "state" (by state we mean some condition of pressure, volume, and temperature), undergoes a thermal process whereby work is done, and then finally is returned to the initial state. In a steam engine, water absorbs heat in the engine's boiler during the process of evaporation to steam. The steam then expands and performs mechanical work. After the work has been performed, the steam is condensed back to water as heat is removed from it in the condenser. Then the cycle is repeated as the water is pumped back into the boiler and converted to steam. This cyclic process is illustrated in Figure 2-9. The importance of going through a cyclic process is that no net change of internal energy of the fluid (or gas) occurs. In other words, the net heat transferred is equal to the work done by the gas.

Since the change in internal energy is zero, we have

$$W = Q_1 - Q_2 \tag{2-6}$$

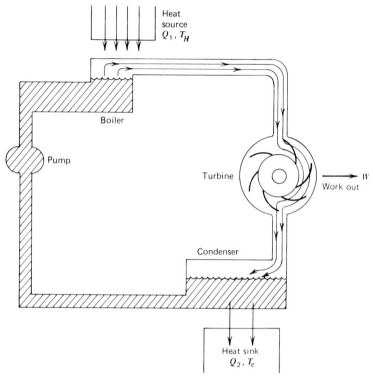

FIGURE 2-9 Illustrates how the fluid in a steam turbine undergoes a cyclic process, returning finally to its initial state. The useful work out is thus equal to the difference between Q_1 and Q_2 (assuming no losses).

which just expresses quantitatively that the work done equals the difference between the heat input to the fluid and the heat delivered to the cooler output. The efficiency of the engine is then

$$\eta = \frac{\text{work output}}{\text{heat input}} = \frac{W}{Q_1} = \frac{Q_1 - Q_2}{Q_1} \qquad (2\text{-}7)$$

Carnot worked out the relationship between heat and temperature for an ideal engine—one with no friction losses—so that the heat level at each end of the cycle would be equal except for the heat energy extracted to perform mechanical work. This concept is important because any real engine has a lower efficiency than the ideal one. Carnot showed that for an ideal engine,

$$\frac{Q_1}{Q_2} = \frac{T_1}{T_2} \qquad (2\text{-}8)$$

where T is the temperature on the Kelvin scale. If we substitute this relation in the expression above for the efficiency, we have

$$\eta_{\text{ideal}} = 1 - \frac{T_2}{T_1} \qquad (2\text{-}9)$$

For an efficient engine the heat source needs to be as hot as possible, and the exhaust temperature should be as low as possible.

The second law of thermodynamics can be stated in many equivalent ways. We have presented it above in the form of a statement about increasing disorder. The essential content of this law is that there is something about the real world that prevents us from extracting heat from a reservoir, converting it to work, and then cooling the reservoir. One of the ways in which the second law can be stated is as follows:

It is impossible to construct an engine, operating in some continuous fashion, which does nothing other than take heat from a source and perform an equivalent amount of work.

REFERENCES

1. Robert H. Romer, *Energy, An Introduction to Physics*, (San Francisco: W. H. Freeman, 1976).

    The appendices in this useful book contain a wealth of information. The author intended that these tables would serve as an "energy handbook." In the opinion of this writer, he has succeeded admirably. Topics covered include conversion factors, physical and chemical data, numerous tables about our various energy sources and uses of energy in the United States.

2. Joseph Priest, *Energy for a Technological Society*, 2nd ed. (Reading, Mass., Addison Wesley, 1979).

    Chapter 3 of this textbook covers some of the same material as this chapter. Priest's book is more quantitative than ours but the coverage is still quite elementary.

QUESTIONS

1. Energy is an important concept in nature because:
    (a) Although its form can change in many ways, it is always conserved.
    (b) It is conveniently related to the square of the force.
    (c) It is simply related to the product of force times velocity.
    (d) The units of energy are the simplest imaginable.

2. A typical unit of energy is the:
    (a) Watt.
    (b) British thermal unit.
    (c) Joule/second.
    (d) Joule-second.

3. Suppose it requires 5.28×10^5 J to lift a weight of 1768 kg from street level to an eighth floor window that is 30.5 m above street level. How many joules are needed to lift this same weight to the sixteenth floor?

    (a) 2.64×10^5
    (b) 5.28×10^5
    (c) 10.56×10^5
    (d) 21.12×10^5

4. Kinetic energy is:

    (a) Gravitational energy.
    (b) Potential energy.
    (c) Energy of motion.
    (d) Nuclear energy.

5. Energy may be defined as:

    (a) A "push or pull".
    (b) The product of force and velocity.
    (c) The capacity to do work.
    (d) The product of mass and acceleration.

6. Which of the following are units of power?

    (a) Kilowatt hours
    (b) Watt
    (c) Horsepower
    (d) British thermal unit
    (e) Kilocalorie
    (f) Joule/second

7. Which of the following cannot be defined in terms of a high-precision atomic standard?

    (a) Time
    (b) Mass
    (c) Frequency
    (d) Length

8. The concept of "energy cost of producing energy" is important because:

    (a) In order to have a practical energy source, the useful energy output must be greater than the energy input.
    (b) The dollar cost of new resources may be too high to be competitive.
    (c) The net energy gain from the development of new resources is always very large.
    (d) The dollar cost of capital equipment in developing an energy resource is so large.

9. Suppose that a lump of coal with an energy content of 1000 J is burned in a coal-fired electric generating plant. About how many joules of electrical energy will be produced by the plant as a result?

(a) 800
(b) 350
(c) 100
(d) 50

10. The energy "cost" of a solar collector for space heating will be made up in how many years of operation?

(a) 20 years
(b) 10 years
(c) 5 years
(d) 1 year

11. Roughly what percentage of the annual consumption of energy resources in the United States goes for space and water heating in buildings?

(a) 1%
(b) 5%
(c) 25%
(d) 50%

12. The Athabasca tar sands in Canada represent a potential oil reserve of about 500 billion bbl. However, less than 20% of this enormous reserve will be recoverable because of:

(a) Gravel and sand in the tar beds.
(b) A very large energy cost to recover the oil from the tar sands.
(c) The necessity of going through a thermodynamic cycle to use the oil produced from the tar sands.
(d) Conservation measures imposed by the Canadian government.

13. The existence of the law of conservation of energy tells us that:

(a) Energy cannot be destroyed and therefore can be "recycled" for use as often as needed.
(b) We need not worry about using up our finite supply of fossil fuels, since the energy is conserved.
(c) The useful energy output from burning a gallon of gasoline cannot exceed its stored chemical energy content.
(d) If oil is burned to generate electricity, all of the energy of the oil consumed can be converted to useful electricity.

14. Circle each of the items below that is a unit of power.

Calorie   Joule-second
Watt   Btu

Kilowatt hour	Btu/hour
Joule/second	Horsepower
Foot-pound	Pound
Kilogram	Newton

15. What happens to the waste energy from a coal-driven electrical generating plant? Is this energy really lost?

16. Suppose that solar radiation is incident on a flat surface at the rate of 400 W/m^2. How much energy is received by 10 m^2 in 1 minute?

17. One common way of heating a home is to run electric current through resistive coils located in the floors and ceilings. Suppose that the electric power is generated in a coal-fired plant. Is this a more efficient way to use the energy of the coal for space heating than to directly burn the coal in a furnace? Explain.

18. Explain why you would expect the fuel consumption per mile of travel by a heavy car to be greater than that of a lighter one.

19. Discuss the arguments for and against the use of waste heat from an electrical generating plant for space heating of residences and commercial buildings.

20. Suppose that roughly one small electric clock per person is in use in the United States. Assume also that each clock requires about 4 W of electric power. If all this electricity were generated by one nuclear reactor, what would be its required power level?

21. One of the following statements about the second law of thermodynamics is not true:
    (a) It allows one to tell how much heat is converted to useful work.
    (b) It is related fundamentally to the conversion of energy between ordered and disordered motion.
    (c) It states that one cannot run any real engine unless there is a finite temperature difference across it.
    (d) It is basically concerned with the definition of the kinetic energy of random motion of molecules.

22. The efficiency of any real engine bears what relationship to that of an ideal engine?
    (a) Greater than   (c) Less than
    (b) Equal to   (d) No relationship

23. Which of the following electric power generating plants has the greatest efficiency of conversion of stored energy into useful electricity?
    (a) A hydroelectric power plant   (c) A coal-fired power plant
    (b) Nuclear reactors   (d) An oil-fired power plant

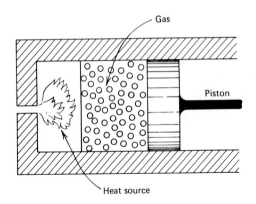

24. Consider the system of heat source, gas, and piston, shown above. Refer to this figure in answering the next two questions. If the piston is held fixed the work done on the piston is:

(a) Zero.

(c) Equal to the change in internal energy of the gas.

(b) Equal to the input heat.

(d) Impossible to calculate.

25. Suppose that when the piston is held fixed, the temperature of the gas increases from T_1 to T_2 when a certain amount Q of heat is added to the gas. Consider next the case where the heat Q is added to the gas, initially at temperature T_1, where the piston is free to move. The gas temperature rises to a new value T_3. Compare T_3 with T_2.

(a) $T_3 < T_2$

(c) $T_3 > T_2$

(b) $T_3 = T_2$

(d) We have insufficient information to compare the two temperatures.

26. What is wrong with using an open refrigerator to cool a house?

(a) The cooling capacity is too low.
(b) Nothing.
(c) The cold reservoir is at too low a temperature.
(d) The hot and cold reservoirs are at the same temperature.

27. In Figure 2-6 suppose that the heat source is replaced with perfectly insulated material. If the piston is now moved inward so that the gas is compressed, what can you say about the temperature of the gas?

28. If the piston in the previous problem had been allowed to move outward, increasing the volume of the gas, what could be said about the temperature of the gas?

29. In which case are the atoms of a lump of coal more ordered—before or after burning? Explain.

30. With what theoretical efficiency can the energy of a stiff breeze be converted to electric power? Is wind an example of ordered or disordered motion?

31. Explain why the change of internal energy in a thermodynamic cycle is zero.

3 nonsolar energy resources

The goal of this text is to help the reader understand how solar energy applications can play an important role in providing for our future energy needs. The solar resource of the United States is extremely large. How does the magnitude of this resource compare with that for our other energy resources? For how long into the future will we be able to rely on oil, natural gas, and coal? In this chapter, the available nonsolar energy resources are described and the problems associated with their utilization discussed.

The literature abounds with widely varying estimates for the energy resources of both the United States and the world. The reasons for this variance can be attributed to both the practical difficulties of estimating reserves and the varying definitions of exactly what should be classified as a reserve. As an example of the latter difficulty, the amount of oil obtainable from oil shale rock could be increased significantly if future technological improvements should permit the utilization of very low-grade shale rock. As an example of the difficulties involved in making reliable estimates, the 1976 U.S. Geological Survey estimates of the reserves of oil and natural gas were revised downward by a factor of 4 to 5 over their 1974 values, mainly as a result of more sophisticated estimation techniques.

Bearing in mind the uncertainties involved in making reliable estimates for the useful energy reserves, let us take a close look at the projected energy resources of the United States. This.will provide us

TABLE 3-1
Estimated Reserves of Depletable Energy Sources for the United States[a]

RESOURCE	PHYSICAL QUANTITY	ENERGY RESERVE (10^{21} J)
Coal[a]	0.4×10^{12} tons	12
Petroleum[b]	74×10^9 bbl	0.40
Natural gas[c]	490×10^{12} ft^3	0.53
Oil shale	600×10^9 bbl	3.54
Fission (^{235}U)[a]	10^6 tons (high-quality ores)	0.53
Breeders (^{238}U)[a]	10^6 tons	74
D-T Fusion	10^6–10^7 tons of lithium	80–800
Steam and hot water		≈ 3.8

[a] *Physics Vade Mecum*, 1981, Chap. 11 (published by the American Institute of Physics.
[b] Using Hubbert's value of 44 billion bbl for the continental United States and adding 30 billion bbl for the Alaska (North Shore) Fields.
[c] 1976 report of the U.S. Geological Survey (50% probability).

with a convenient point of reference for all further discussion. It is useful to make a division of energy sources into those that are depletable (or nonrenewable) and those that are renewable, that is, continually usable. Table 3-1 lists reasonable 1980 estimates of U.S. nonrenewable energy reserves. Noting that the 1980 U.S. annual energy consumption was about 8.2×10^{19} J, we see that if oil were used to provide for all of the energy needs of the country, there would only be a five-year supply left. However, it is impossible to extract the oil from the ground and deliver it for use at this rapid a rate. In 1981, with oil providing about one-half of the country's energy needs, the amount of imported oil roughly equalled that produced domestically.

One could argue, perhaps correctly, that the oil shale and uranium fission resources are considerably larger than shown because of the availability of low-grade ores that will some day prove useful. In the case of oil shale, account has not been taken of the energy required to obtain the oil from the rock. The energy potential of nuclear fusion obtained from the reaction of deuterium with tritium is seen to be quite large, although the amount of basic raw material available is quite uncertain. At present, the extraction of useful energy from fusion processes is still a scientific dream.

Table 3-2 lists the more reliable estimates of the U.S. renewable energy resources. The numbers in the last column indicate the fraction of the 1980 U.S. energy consumption that each resource could supply for an indefinite period if it were fully developed. No account has been taken here of practical losses associated with converting the renewable resource into a useful form. All of these, except for tidal power, are just

TABLE 3-2
Renewable Energy Resources for the United States

RESOURCE	AMOUNT OF ENERGY (10^{21} J)	FRACTION OF 1980 ENERGY USE
Solar radiation	46.8	570
Wind power	0.32	3.9
Ocean thermal gradients	0.38	4.6
Hydroelectric	0.008	0.10
Photosynthesis	0.015	0.18
Organic wastes	0.006	0.07
Tidal energy	0.006	0.07

different facets of solar energy. It is possible that development of photosynthesis in the oceans, such as through the growth of kelp beds, could significantly increase the estimate for this item.

A. OIL AND NATURAL GAS

Estimates of the amount of oil and natural gas that will ultimately be discovered and produced have differed dramatically over the years. One reason for this is that oil and gas deposits occur in a restricted volume of space and in limited areas in sedimentary basins at all depths from several hundred to several thousand feet, so that reliable yield predictions are difficult. In addition, estimates were often made using improper techniques. In 1956 the renowned geophysicist M. K. Hubbert presented his estimate of the available U.S. reserves, as shown in Figure 3-1, using a sophisticated method of analysis. These early conclusions of Hubbert have proven remarkably accurate over the years. In 1961 Hubbert made an improved study that predicted that the ultimate continental U.S. production would be 170 billion bbl with a peak in the rate of production at about 1968, with an uncertainty of not more than two or three years. Later studies by Hubbert in 1971 and 1977 have validated these conclusions.

Within a few years after Hubbert's conservative predictions of about 150 to 200 billion bbl of oil as the ultimate U.S. productive capacity, the petroleum industry came up with estimates of the total capacity as high as 600 billion bbl. In 1961 the U.S. Geological Survey presented its estimate of total production as 590 billion bbl of oil. If true, there would have been no expectation of an oil shortage before the year 2000.

How were these higher estimates arrived at? Essentially they followed from the assumption that the amount of oil discoveries per foot of exploratory drillings would remain constant. Therefore, by assuming

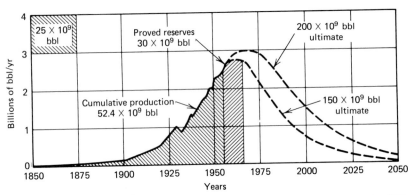

FIGURE 3-1 The U.S. oil production cycle, according to M. K. Hubbert in 1956. Estimates are shown for two different values of the total U.S. oil capacity. [Originally published in the *Canadian Mining and Metallurgical Bulletin*, Vol. 66, No. 735, pp. 37–54 (1973)].

that only 20% of the oil exploration had been completed in 1959, at which time the total cumulative production was about 100 billion bbl, it follows that the ultimate production capacity would be of the order of 500 billion bbl. The fallacy in this argument is that new oil is becoming much more difficult to find. In the 1930s the oil industry found 275 bbl of oil for each foot of exploratory drilling, whereas in recent years this has fallen to less than 35 bbl per foot.

An important aspect of the bell-shaped production curve shown in Figure 3-1 is the insensitivity of the year of peak production to the ultimate reserve (called Q_∞). Changing Q_∞ from 150 to 200 billion bbl only changes the year of peak production from 1966 to 1971. To drastically change the year of peak production to the year 2000 AD would require that Q_∞ be about 600 billion bbl, a fourfold increase. In fact, the U.S. production of oil peaked near the year 1968 and has been decreasing since, despite massive efforts to increase production.

In the face of decreasing production at home, the United States is becoming increasingly dependent on imported oil, mainly from the Middle East. The principal oil reserves in this portion of the world lie in Saudi Arabia (130 billion bbl) and in Iran and Kuwait (60 billion bbl each). In 1960 two eminent energy authorities were quoted as saying, "Imports involve no foreseeable problem of supply, either in availability at the source or in transportation." Because of the political problems in the Middle East and the 1973 Arab boycott, this position would not appear to be particularly reasonable now.

Although the oil situation is serious, our cleanest fuel, natural gas, may be the first to give out. The 1962 estimate of M. K. Hubbert is shown in Figure 3-2, which implies that the U.S. reserves in 1975 would be about 400 to 500 trillion ft^3. This estimate is in good

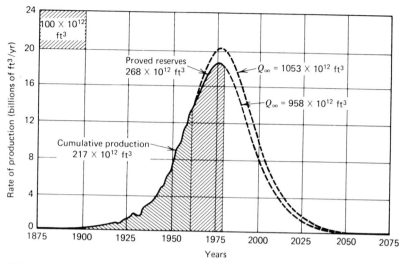

FIGURE 3-2   1962 estimates of M. K. Hubbert of the U. S. natural gas reserves. His prediction of the amount remaining after 1975 is in remarkable agreement with the U.S. Geological Survey estimate. [Originally published in the *Canadian Mining and Metallurgical Bulletin*, Vol. 66, No. 735, pp. 37–54 (1973)].

agreement with the latest U.S. Geological Survey estimates of 320 to 655 trillion ft^3, with the same confidence percentages for the low and high estimates as for oil. Since 1968, the United States has been using natural gas at a rate that is twice that of new discoveries. Many localities will no longer service new consumers. Imports of natural gas using large, refrigerated tankers that cool the gas to $-260°F$ and reduce its volume by a factor of 600 have helped to temporarily alleviate the shortage.

From Table 3-1, it is seen that shale rock contains large amounts of oil. The Department of the Interior has estimated that the Green River formation in Colorado, Utah, and Wyoming, shown in Figure 3-3, contains about 600 billion bbl of oil—20 times the Alaskan reserves and about 12 times as much as the established conventional oil deposits in the country. Actually it has been estimated that the Green River shale deposits contain up to 2 trillion bbl of oil, but only about one-third of this is in reasonably thick deposits that average more than 25 gal of oil per ton of shale; only these are presently regarded as commercially exploitable.

Extracting the oil from oil shale rock is not a simple process, even though the technological details of the extraction process are straightforward. When shale is crushed and heated to 480°C, raw shale oil is released. But this process is energy intensive, and the energy cost of producing this oil from shale rock must be carefully evaluated. Furthermore, large amounts of water are needed; the production of 1 million

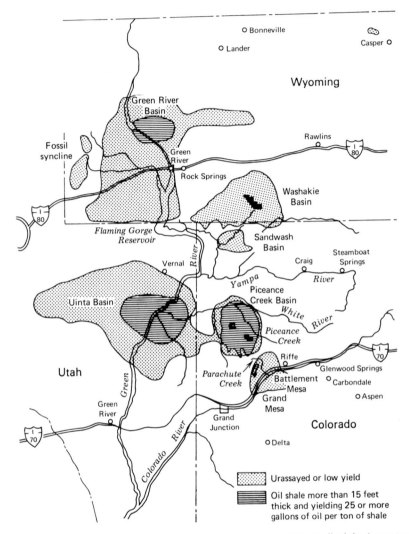

FIGURE 3-3 Map of the major U.S. oil shale deposits. Almost all of the important reserves that are probably recoverable occur in this Green River area. [From *Science*, 184, 1272 (June 1974).]

bbl a day requires about 3 million bbl of water although recent technological developments may drastically reduce this requirement. By a bitter twist, the shale is located in some of the driest areas in the United States. The primary water supply is the Colorado River and every drop taken increases its salinity.

An exciting development is the possibility of converting the kerogen in the oil shale into crude oil in situ. If successful, this technology not only would reduce the cost of shale oil but also would greatly reduce

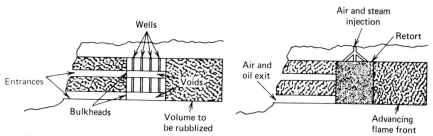

FIGURE 3-4   One suggested approach for the extraction of oil from shale rock in situ. A description of the process is given in the text. [From *Science*, 198, 1203 (December 1977).]

the amount of water required and alleviate the problem of disposing of the spent shale. A number of approaches of in situ processing have been tried. Figure 3-4 illustrates the scheme evaluated by Occidental Petroleum Company. In this approach about 20% of the shale rock is excavated and relocated on the surface by conventional means. The remainder of the shale is then fractured with explosives, and the shale rubble expands to fill the voids. The shale is then ignited and air and steam are injected at the ceiling. The oil from the shale flows downward ahead of the flame front and collects in a sump.

B.   COAL

Estimation of the amount of coal reserves in the United States and the world can be done much more reliably than was the case for oil. This is because coal is found in stratified beds or seams that are continuous over extensive areas, and geological mapping can be done on the basis of a few widely spaced drill holes. On this basis, U.S. geological reserves are estimated to lie between 1.0 to 2.0×10^{12} tons, which is about one-fifth to one-third of the world's supply of recoverable, bituminous coal and lignite. This amount is about 3000 times the 600 million tons of coal consumed in 1970. Our realizable coal reserves may be considerably smaller than this, however, since only a fraction of the total resource can be considered minable under present economic and technological conditions. It may be that only 200 to 400 billion tons of coal should be considered as a proven reserve for the United States (see Table 3-1). Even so, coal should be a significant factor in the energy equation for the next 200 to 300 years.

Coal production today is about equally divided between underground and surface mining. An ever increasing amount of coal is obtained by strip mining, even though only about 3 to 5% of the total coal resources can be mined in this manner. The large increase in strip mining has resulted primarily from the introduction of highly efficient, heavy excavating equipment. Mammoth drag lines with bucket capacities up to 168 m², power shovels, and bucket excavators weighing nearly 8000 tons are now in use. This method enjoys a real cost advantage over deep-mined coal, but so far has left the country with many examples of lunar landscapes, such as that shown in Figure 3-5. Strip-mined areas can usually be reclaimed, and the cost is not prohibitive, but tougher reclamation laws will be needed.

Underground mining of coal also presents many special problems. Dangerous elements such as lung-blackening dusts and explosive methane gas make underground coal mining a risky profession at best. The environmental impacts from deep mining—acid mine drainage, land surface subsidence, gas disposal, and mine and waste heap fires—are far from trivial. Underground mining is generally more wasteful of the coal resource than surface mining; only 57% of the minable coal is presently recovered with the rest of the original resource left in the mine, usually rendered unfit for future recovery.

A serious drawback of coal is its sulfur content, resulting in a large amount of sulfur dioxide (SO_2) after burning. This is commonly removed from the stack gases after combustion. But, a dependable method has not yet been found to accomplish this economically.

FIGURE 3-5 Typical lunar-appearing landscape from strip mining in North Dakota. This unreclaimed land was mined years ago. Volunteer native shrubs and trees have established themselves in the valleys, and grass sown by the wind covers the windward side of the slopes (the left-hand side of the hills in the picture). No company leaves land in this condition today without violating both state and federal laws. (Courtesy of National Coal Association.)

Several once promising techniques to remove the pollutants from stack gases have had to be abandoned and others are running into trouble. Research scientists and engineers are now looking at schemes where the pollutants are removed at an earlier stage in the combustion process. Historically, such a process was used near the turn of the century to provide clean heat at high temperatures for such purposes as heat-treating metals or producing ceramics and fine glassware where the dirty products of coal combustion could not be used. A two-step combustion process was devised. Instead of supplying air to a shallow bed of coal and burning the coal to convert its carbon and hydrogen into carbon dioxide (CO_2) and water vapor, air and steam were blown through a deep bed of coals to obtain a fuel gas composed mainly of carbon monoxide (CO), hydrogen (H_2), and nitrogen (N_2). This gas, after cooling and scrubbing with water to remove dust, was burned to provide the desired clean heat. For electricity production, this clean "power gas" can be burned in a gas turbine. Technological improvements in gas-turbine design over the years have resulted in steady increases in both efficiency and size; 300-MW generators are now contemplated. The hot gases rejected from the gas turbine can also be used to drive a conventional steam turbine, with overall efficiencies of over 50% in prospect for the combined two-stage, gas-steam generator system.

Considerations such as the above, combined with the need to use more coal in the future, furnish a considerable economic incentive for coal gasification schemes. One such system is already in commercial operation, the coal-gas process of the Lurgi Gesellschaft für Mineralöltechnik Company in West Germany. This system is imperfect, however, since it results in an undesirable release of water vapor, rapid cooling of the power gas, an inability to work on small pellets of coal, and a limited coal-processing capability. Because of the importance of coal gasification, it can be expected that significant improvements will be made in the next few years.

C. AIR POLLUTION

An increase in the use of materials and energy in the United States has resulted in an ever-rising rate of pollution. The results of this pollution have now reached the point where the adverse environmental effects are noticeable on a large scale. David Freeman, formerly Chief of the Energy Policy Staff at the White House, has stated the situation quite clearly and succinctly, "Americans as a nation no longer feel that we can produce and use energy with total disregard for its effect on the environment."

The many aspects of pollution and its control form the basis of entire

TABLE 3-3
Estimated Emissions of Air Pollutants by Weight[a]

POLLUTANT	CO	PARTICULATE	SO_x	HC	NO_x	TOTAL
Auto	82.1	1.3	0.8	12.5	10.6	107.3
Power plants	1.4	7.0	26.8	1.7	13.3	50.2
Industrial	11.0	10.6	6.3	3.7	0.7	32.3
Refuse burning	3.5	0.6	0.1	1.0	0.2	5.4
Miscellaneous	5.3	0.8	0.1	14.0	0.2	20.4

[a] Units are millions of tons per year. These data are from *National Air Quality and Emission Trends Report, 1975*, U.S. Environmental Protection Agency Report Number 450/1-76-002.

books. All that will be covered in this section is a quick survey of some of the key pollution problems associated with the use of fossil fuels. As noted earlier, one serious problem is the resultant water and ground contamination from oil and coal production. Another serious aspect is the air pollution resulting from the consumption of fossil fuels to perform useful work. We are perhaps most acquainted with air pollution through the unhappy phenomenon of smog, a collection of irritating hydrocarbons and oxides of sulfur and nitrogen. The dominant air pollutants in the United States are listed in Table 3-3.

Since the burning of coal, particularly to produce electricity, is liable to be the dominant source of energy in the United States over the next century, it is worthwhile to emphasize the serious nature of the environmental concerns that result. In a large coal-burning power plant, the principal emission from the plant's exhaust stacks is carbon dioxide (not shown in Table 3-3). Although carbon dioxide is not a serious pollutant in itself, there is growing concern that the dumping of vast amounts into the earth's atmosphere could seriously affect the planet's climate. The most harmful pollutants released are oxides of sulfur. Large emissions of these sulfur compounds can cause deaths from cancer, damage to human lungs, and property damage. Other dangerous emissions are nitrogen oxides, the principal pollutants in automobile exhausts and fine particulate matter.

Pollution can be controlled. Emissions from the 950-MW steam plant at Bull Run near Oak Ridge, Tennessee are relatively low. A giant bank of precipitators traps most of the fly ash, about 950 tons/day. This can be contrasted with plants near Chicago and the Four Corners area of New Mexico, where the fly ash pours out, although recent efforts in New Mexico have resulted in an order of magnitude improvement.

Not all pollution control need be performed at the source. One of the better ways to minimize the adverse effects of urban industrialization is through the use of green areas such as golf courses, parks, and so on. It has been found that a park or golf course, plentifully covered with trees, can effect a 10 to 25% reduction in air pollutants such as sulfur dioxide (SO_2), oxides of nitrogen (NO), and hydrogen sulfide (H_2S), in the neighboring area.

Perhaps the most attractive feature of utilizing solar energy is that it is essentially a nonpolluting energy resource. Space heating of homes by solar energy is an ideal example of this. A solar collector system gathers the incoming insolation, stores it as thermal energy, and distributes it to the house, with a resultant leakage to the outside atmosphere. There is practically no change in the heat balance of a city between the situation with and without solar-heated buildings. No pollutants are emitted from the house as is the case for gas, oil, wood, and coal furnaces. It should be noted that a rapid buildup of solar heating systems, with the attendant manufacturing of collectors and other hardware, will result in extra pollution. But after these systems are in place pollution will be abated. (For electric heating the pollutants are emitted at the generating plant.) The energy and economic cost of a solar space heating system is not much greater than that of a conventional heating system. The impact on the environment of solar electric plants is harder to evaluate, since all of these are in very early stages of development; but they should be relatively nonpolluting by comparison with other energy sources.

D. GEOTHERMAL ENERGY

Geothermal heat comes from hot rock in the earth's interior, which is at temperatures near 500 to 1100°C. As it is conducted to points near the surface, ground water is heated to temperatures near 200 to 400°C. The resulting geothermal energy resources can be divided into four distinct types.

Dry steam wells are formed when the groundwater deep within the earth has a high temperature at low pressure resulting in underground boiling. The accumulated steam can be tapped by wells ranging in depth from 600 to 2400 m, releasing the dry steam directly. The hot steam is then run through centrifugal separators to remove the rock and grit. From these it is fed directly into low-pressure steam turbines to produce electricity. Electricity is now produced from dry geothermal steam fields in Larderello, Italy (384 MW); the Geysers Field, Sonoma County, California (more than 600 MW); and at the Matsukowa Field in Japan (50 MW).

Wet steam geothermal areas are much more abundant than dry steam areas. In this case, hot water in the earth is kept under high pressure, and boiling does not occur underground. When tapped, the superheated water flows into the well and changes into a vapor. The steam must be separated from hot water before it can be used to generate electricity.

The waste hot water is allowed to run off and, since it contains many pollutants, serious environmental problems are involved.

Hot water fields at 65 to 120°C are primarily used for heating or refrigeration. They can also be used to generate electricity by boiling a secondary fluid such as freon, with a low boiling point. The secondary fluid is then used to drive power turbines, but the energy conversion efficiency is very low. Small plants of this type are being used in New Zealand, in Kamchoika in the USSR, and in Japan. The city of Reykjavik, Iceland, is heated geothermally, with hot water used for domestic purposes, swimming pools, and so on. Using the hot water for heating in this manner may prove the best way to tap the energy of this large geothermal resource.

A program to tap the geothermal energy contained in hot rocks at depths of more than 4000 m is being pursued by the Department of Energy. Pressurized water is used to "fracture" the hot rock, as in oil drilling. Water is brought down to this depth through one pipe where it turns to steam. This steam is then piped up through another hole that has been drilled close by and flows into a turbine generator where electricity is produced. The hot rock fractures continously and conceivably such a well would last about 30 years, producing power on the order of 50 to 100 MW.

E. NUCLEAR FISSION

Of all the new sources of energy, nuclear fission has received the most support, and its technology is the most well developed. Over 70 nuclear reactors are now operational in the United States. Until recently it has been commonly thought that nuclear reactors would rapidly take over from fossil fuels the lead role in the generation of electricity. While this may still come about, there is a growing concern in both the general public and the scientific community about the extent to which research funds have been concentrated on this energy option and about the wisdom of basing an energy economy on nuclear power. Increasing awareness of the possible consequences of the large-scale use of nuclear fission as a source of energy has also contributed to this reevaluation of our energy priorities. The debate over the pros and cons of nuclear power has often risen to a white-hot fury. The actual determination of a national policy on this subject is still to come. Nevertheless, the nuclear age will undoubtedly be with us for a long time.

The emphasis in this section is on the potential of conventional nuclear reactors as an energy source. The essential nuclear physics needed to understand the fission process is well covered in the many fine books on this topic. It is sufficient for our present discussion to note that when a slow neutron interacts with a fissionable isotope such as ^{235}U (this is uranium with 92 protons and 143 neutrons. Natural uranium also has an isotope, ^{238}U, with 92 protons and 146 neutrons), the resultant object will split into two almost equally massive fragments, with a consequent large release of energy. Energy is also released when two light nuclei are combined together in a fusion reactor. The possibility of obtaining useful energy from fusion reactions is covered in the following section of this chapter.

A nuclear reactor is based on the principle that the fission processes occur in an environment such that the number of neutrons given off during fission are just exactly sufficient to maintain the process. This means that if 2.5 fast neutrons are emitted, on the average, during a typical fission event, then exactly 1.0 neutrons must remain, on the average, after the emitted neutrons have undergone a typical life cycle.

Most nuclear reactors in the United States have a number of common features. The fissioning material, ^{235}U, is formed into long slender rods spaced uniformly throughout the core of the reactor. (Most of the uranium in the fuel rods is ^{238}U; only about 4% of the uranium is ^{235}U.) The neutrons emitted in the fission process come off with very large velocities and cannot cause ^{235}U to fission. This necessitates filling the reactor volume between the fuel rods with a light material called the moderator, which serves the purpose of slowing down the fast neutrons. Because of the fissioning material, the fuel rods become hot and a coolant, usually water, is passed over them in order to remove the energy generated. To extract useful power the hot coolant is then passed through some type of turbine where electric power is generated. To maintain the reactor in a stable operating condition and furnish protection in the case of the failure of any of the reactor components, control rods that readily absorb neutrons are needed. Protection against a loss of water is provided by an emergency cooling system that is designed to force cool water, at high pressure, into the reactor core.

It is instructive to examine the life cycle of a typical neutron emitted from a fission event. The average history of such a neutron is depicted in Figure 3-6. Some of the fast neutrons created in the fission process either leak out of the reactor volume or are absorbed by the uranium (mostly ^{238}U) in nonproductive processes. This happens during the time period when the moderator is slowing down the fast neutrons. After this slowing-down process, about 1.7 of the original neutrons are left (on the average, of course). Of these slow neutrons, 0.5 are absorbed by material other than ^{235}U, and 0.2 leak from the reactor volume. In the stable situation exactly 1.000 ... neutrons are left over to instigate a new fission reaction.

It only takes a small fraction of a second for this neutron cycle to be completed. If all of the neutrons in the reactor went through exactly this cycle, control of the power level of a nuclear reactor would be

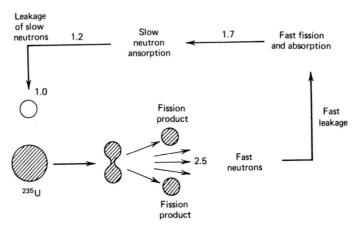

FIGURE 3-6   Schematic illustration of the life cycle of neutrons in a typical reactor situation. At equilibrium, the original 2.5 neutrons are just sufficient to overcome leakage and unwanted absorption processes, providing one slow neutron to initiate another fission event.

impossible. Fortunately, about 1% of the emitted neutrons arise from processes that are very much delayed from the original fission event. Because of this fortuitous circumstance, reactors can be controlled by the relatively slow movements of control rods.

Natural uranium is a plentiful element, even being found in small amounts in granite and in sea water. However, only about 0.7% of the natural uranium consists of fissionable ^{235}U, the rest consisting of ^{238}U, which is an isotope containing three more neutrons and which is not fissionable by slow neutrons. Because of this circumstance, there is a lower limit to the concentration of uranium in naturally occurring ores that can usefully be exploited. An estimate of the useful uranium reserves in the United States as published in a 1973 report by the old Atomic Energy Commission is given in Table 3-4.

The reserves are listed in terms of tons of uranium oxide (U_3O_8). For example, the first row shows that there are about 1.13 million tons of uranium oxide available with a naturally occurring concentration greater than or equal to 1600 parts per million (pure uranium ore). There are an additional 1.63 million tons of ore with a concentration ranging from 1000 to 1600 parts per million, and so on. One sees immediately that the fission reserve for conventional reactors is rather small if only the low-cost, almost pure uranium ore reserves are used. In fact the energy potential from this fraction of the uranium reserve is hardly any greater than that from the remaining U.S. oil reserves if only the first two rows of Table 3-4 are included. Making use of the lower-grade ores will alleviate this situation somewhat. However, the real attractiveness of nuclear power for the long term lies with the potential

TABLE 3-4
U.S. Uranium Reserves

U_3O_8 CONCENTRATION (*PARTS PER MILLION)	AVAILABLE U_3O_8	ELECTRICAL ENERGY[a] (10^{21} J)	
		CONVENTIONAL REACTORS	BREEDER REACTORS
1600	1,127	0.21	28.2
1000	1,630	0.30	40.6
200	2,400	0.44	59.5
60	8,400	1.56	208
25	17,400	3.26	431

[a] Includes an overall efficiency of 32% for the conversion to electrical power.

of the breeder reactor. This approach permits more fissionable material to be manufactured than is consumed. This occurs because ^{238}U, the dominant uranium isotope, can be converted into ^{239}Pu (with 93 protons and 146 neutrons) when it interacts with a fast neutron. And ^{239}Pu is a fissionable isotope, just like ^{235}U.

A reactor cannot be constructed entirely of natural uranium because the much more plentiful ^{238}U absorbs too many neutrons to permit a chain reaction to take place. However, a chain reaction can occur in a large lattice-type structure composed of uranium metal interspersed with a lightweight moderator such as ^{12}C (with 6 protons and 6 neutrons), water, or "heavy water" (D_2O). The purpose of the moderator is to slow down the fast neutrons as quickly as possible in order to avoid their being captured in nonfission processes. This is accomplished by using the fact that whenever a fast neutron collides with a light atom it loses energy and hence has a smaller velocity; the lighter the struck nucleus the more energy the neutron loses. On the average the maximum energy transfer occurs when the struck nucleus has the same mass as the neutron. From this standpoint the hydrogen in ordinary water would be the best moderator; unfortunately, part of the time hydrogen captures a neutron to form deuterium (heavy hydrogen), which is an additional absorption process that does not lead to useful fission. Consequently, ordinary water cannot be used with natural uranium as the fuel, but instead it requires uranium that has been enriched in the content of ^{235}U.

For reasons that were partly economic and partly political, industry in the United States has chosen to go with ordinary light water as the coolant and moderator. This water is circulated under pressure thereby allowing elevated temperatures up to 330°C to be achieved. There are two types of light-water reactor that are now in use. In one, the pressurized reactor, used in about 60% of the nuclear power installations, the water coolant is maintained under high pressure [150

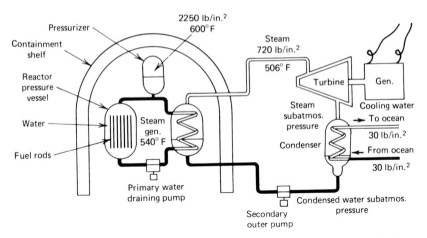

FIGURE 3-7 Illustration of the basic components of a conventional pressurized, water fission reactor. (From *Perspectives on Energy*, C. Ruedisili and M. W. Firebaugh, eds., Oxford University Press, New York, 1975; with the permission of the National Academy of Science, Washington, D.C.)

atmospheres (atm)] to allow high coolant temperatures without steam formation within the reactor. The hot coolant then passes through a heat exchanger where water is converted into steam. Steam from this system is channeled to a turbine driving an electric generator. A schematic diagram of a pressurized-water reactor is given in Figure 3-7. The other commercial installations are known as boiling-water reactors. In this case the water in the reactor is held at a lower pressure and is allowed to boil inside the reactor core. The steam and water flowing past the fuel are then separated with the steam flowing directly to a turbine. Both types of reactors have thermal efficiencies (electric power output per energy input) of about 32%, compared to a typical value of 40% for a fossil-fueled plant that can operate at higher temperatures.

Since U.S. reactors utilize light water as the moderator, they must then enrich the percentage of ^{235}U in naturally occurring uranium. This is presently done at Oak Ridge, Tennessee, where solid U_3O_8 is converted to gaseous uranium hexafluoride (UF_6) and then passed many times through a gaseous diffusion process. The uranium fuel, enriched to 4% ^{235}U, is then suitable as the fuel for the light-water reactors. A large fraction of the energy cost for the nuclear power industry occurs in this step. In order to enrich the ^{235}U content to 4% for the desirable fuel, 25% of the original ^{235}U is left over as reject material in the form of depleted uranium "tailings" made up of uranium with a ^{235}U content of 0.2%. Because of the need to enrich, it takes about 5 tons of pure uranium ore to give 1 ton of enriched fuel. A typical 1000-MW reactor uses about 30 tons of enriched fuel in one year. Present reactor practice is to consume the ^{235}U in the reactor until the ^{235}U percentage is down to about 1%.

Ordinary water is not the only choice for a moderator. The original Hanford reactors, designed to produce the fissionable isotope ^{239}Pu during the Second World War, used very pure graphite as the moderator. Heavy water is used as the moderator in the Canadian reactor program. This permits these CANDU reactors (for Canadian-deuterium) to use natural uranium as the fuel. Since no enrichment is needed with its subsequent rejected material, and since almost complete burnup of the ^{235}U is possible in this approach, the CANDU reactors can get about twice as much energy from a ton of natural uranium as does the typical light-water U.S. reactor under present operating conditions, when no reprocessing of the spent fuel is done.

In many ways nuclear fission has substantial advantages over traditional sources of energy. Air pollution from the consumption of fossil fuels is still a problem, whereas nuclear power plants do not emit particulates, sulfur oxides, or other combustion products. Nuclear fuels are a compact source of energy, which has advantages in terms of size and location of power plants. If the breeder reactor proves to be a valid future energy source, then nuclear power can provide for the world's energy needs for perhaps several thousands of years. Nevertheless, society has not yet generally accepted nuclear power as the best solution to the world's energy problems. The major concerns about nuclear power can be grouped roughly into the following categories: (1) routine emission of radioactivity from the reactor, (2) radioactive waste disposal, (3) safety problems such as the possibility of a loss of coolant accident, (4) disposal of radioactive tailings from the mining and enrichment processes, and (5) the problem of safeguarding an ever-increasing inventory of fissionable material. Other problems such as thermal pollution from waste heat are common to all electric power generation schemes.

F. NUCLEAR FUSION

When two very light nuclei combine, energy is released. The most spectacular example of this type of reaction occurs in the sun where hydrogen nuclei fuse together to form helium (see Chapter 5). This process cannot be utilized on earth because the probabilty that two hydrogen nuclei will combine when they collide with each other is extremely small. A large mass of hydrogen, such as that present in the core of the sun, is needed. Fortunately, other fusion reactions are available that offer the possibility of practical fusion power here on earth. The search for practical fusion power is being enthusiastically pushed because the rewards, if the quest proves successful, are large.

The most promising fusion reactions involve the heavy hydrogen isotopes deuterium and tritium. Deuterium is a naturally occurring isotope while tritium, which decays radioactively, must be manufactured in some manner. The important fusion reactions involving these isotopes are the following:

$$D + D \rightarrow {}^3He + n + 3.2 \text{ MeV} \qquad (3\text{-}1)$$

$$D + D \rightarrow T + n + 4.0 \text{ MeV} \qquad (3\text{-}2)$$

$$D + T \rightarrow {}^4He + n + 17.6 \text{ MeV} \qquad (3\text{-}3)$$

The energy released is written in million electron volts (MeV) for simplicity. For comparison with our regular energy units note that 1 MeV equal 1.6×10^{-13} J.

The most attractive feature of the first two reactions is the almost inexhaustible supply of deuterium available from the oceans. One out of every 6500 atoms of hydrogen in water is the heavy isotope deuterium. Thus, a cubic mile of sea water is equivalent to about 8.3 trillion barrels of oil, a quantity considerably greater than the total world reserves of oil. For all practical purposes it is accurate to say that the fusion of deuterium represents an essentially infinite supply of energy.

In the case of fusion, there is a difficulty that arises which was not present in the fission process where we were dealing with the capture of neutrons. Neutrons are electrically neutral while deuterium and tritium are charged particles. Two similarly charged particles are repelled by the repulsive electrical force which increases the closer the charged particles come. This repulsion can be described as an energy barrier, with the energy needed to get over the barrier as shown plotted in Figure 3-8. It is the electrical charges on the reacting particles that are responsible for this barrier. In order for a useful fusion reaction to take place this barrier must somehow be overcome. When the interacting particles are separated by less than 2×10^{-13} cm the attractive nuclear forces predominate and a nuclear reaction may occur. In order to get this close the interacting particles must have sufficient speed to overcome the barrier. It is exactly like having a car move very fast as it approaches a hill, then turning off the engine and seeing how far up it will coast. With sufficient initial velocity it will coast over the top. In the case of our elementary fusion nuclei this is accomplished by heating up a gaseous mixture to extraordinarily high temperatures, of the order of 10^8 K.

Once the two charged particles have overcome the repulsive electrical barrier, there is still only a small probability that they will interact in the fashion indicated by Eqs. 3-1, 3-2, and 3-3. In fact, the reaction process of Eq. 3-3 will take place much more readily than those involving the interaction of two deuterium nuclei. Because of this, present efforts to achieve useful nuclear fusion energy have concentrated on the D-T reaction. Inasmuch as tritium is not a naturally occurring isotope, it will be necessary to produce it from another reaction. We can

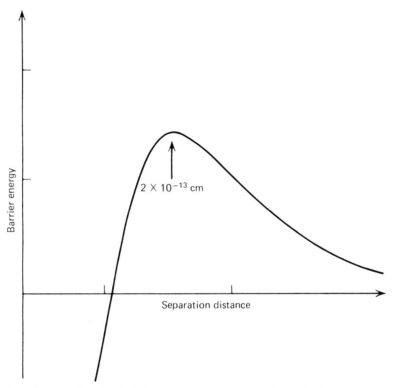

FIGURE 3-8   The electrical force between two charged particles increases as the separation distance decreases. This is represented as an energy barrier which must be overcome. At a separation distance of about 2×10^{-13}/cm the attractive nuclear force takes over and a fusion reaction can take place.

do this by allowing the fast neutrons from the D-T reaction to interact with natural lithium.

To produce net fusion power, a gaseous mixture of deuterium and tritium must be heated to the combustion temperature for fusion and then must remain in this high-temperature condition long enough for the recovered output of fusion energy to exceed, on the average, the energy brought into the system. Lawson has developed a crude way of stating this requirement in terms of the plasma density, n, and particle confinement time, τ. This condition is obtained by balancing the net fusion energy recovered during the confinement time τ against the energy of motion of the high-temperature plasma—a charged particle gas composed of an equal mixture of positively charged nuclei and free electrons. The result is often plotted in the form shown in Figure 3-9. That the Lawson criterion can be stated in the form of a product, $n\tau$, follows from the fact that when the plasma density is increased the rate of fusion burning is correspondingly faster, so that the confinement

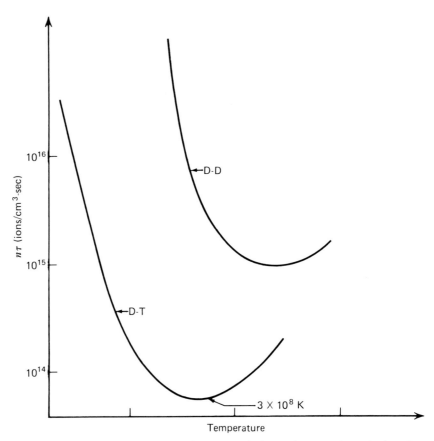

FIGURE 3-9   The product of plasma density n with the confinement time τ is plotted as a function of the temperature of the plasma particles. The minimum $n\tau$ product for D-D fusion is over an order of magnitude larger than that for the D-T fusion reaction.

time needed to reach the break-even energy point is shortened and vice versa when the plasma density is decreased. For the D-T reaction, at a particle density of 5×10^{14} (cm^3), the break-even confinement time is about 0.2 sec.

From the above discussion we can see that the crucial problems to solve in order to obtain practicable fusion power involve the heating of the plasma and its confinement in a hot, dense state. In the past 30 years of fusion research two quite different approaches have evolved. The first one involves magnetic containment of the plasma in which obtaining a sufficiently long confinement time is the primary problem. The second approach involves pellet fusion, where the key aspect is heating. The Lawson criterion is universal in that it applies to any fusion situation. For magnetic containment systems, confinement times that are appreciable fractions of a second are needed. On the other hand, for pellet fusion, the particle densities are thousands of times the density of ordinary matter and τ need only be of the order of 10^{-11} sec. At the present time, magnetic containment approaches appear to be the most promising.

A central aspect of magnetic containment schemes concerns the existence of plasma instabilities. These instabilities can lead either to rapid expulsion of the plasma from the magnetic field or to slower, but nevertheless undesirable, diffusion out of the containment geometry. Thanks to a large amount of innovative research, the instability problems have been overcome through the use of technically complicated magnetic field configurations.

How will energy finally be extracted in a useful way? The charged particles will be contained within the plasma, but not the neutrons which will come off with about 75% of the kinetic energy released in the reaction. A way is needed to make use of this kinetic energy. Two suggestions appear feasible. The neutrons can be used to heat up a moderator such as water, generating steam to drive a turbine which will then produce electricity. Or the neutron's kinetic energy can be used to heat up molten lithium, whose heat would then be transferred to water, producing steam and giving electricity in a customary fashion. The latter process has an additional advantage in that it will help produce tritium through the reaction

$$^{6}Li + n \rightarrow T + {}^{4}He$$

The design of a practical fusion reactor will present many difficult engineering problems because enormous differentials of temperature and neutron flux must be sustained over very small distances—the temperature at the center will be roughly 10^{8} K and the neutron flux will be enormous. Pumping liquid lithium through the system will be troublesome and the lithium must be protected against contact with water; otherwise a dangerous fire could result. Because the supply of lithium in the world is quite limited, the total energy available from this resource appears to be about the same as that obtainable from fossil fuels. It is the more difficult D-D fusion using the deuterium from the oceans which is often referred to as the ultimate energy source. Conceptual studies of magnetic containment reactors indicate that they will be considerably larger than fission plants as for a given power level, and in many ways more technically complex.

Progress with fusion research has not yet reached the point where the same amount of energy is taken out as is put into the device. For a useful fusion reactor, the energy output must greatly exceed the energy input. Despite an enormous research effort, we have not reached the stage where fusion energy is technically feasible. Clearly, the great adventure of obtaining useful energy from fusion reactions on earth is just beginning.

REFERENCES

1. L. Ruedisili and M. Firebaugh, eds., *Perspectives on Energy* (New York: Oxford Univ. Press, 1975).

    One of the very best books on energy. This volume is a collection of essays by experts on the basic issues as well as on fossil fuels, nuclear fission, and other energy alternatives. The article by M. K. Hubbert that surveys the world's energy resources is especially recommended. We also highly recommend a review by Hubbert published in the November, 1981 issue of the *American Journal of Physics* (pp. 1007–1029).

2. Robert H. Romer, *Energy, An Introduction to Physics* (San Francisco: W. H. Freeman, 1976).

    Appendices H, I, K, and L of this general physics text contain a wealth of valuable data about the energy content of fuels, about production and consumption of energy, and about our sources and uses of energy.

3. Jack M. Hollander and Melvin K. Simmons, eds., *Annual Review of Energy*, Vol. 1 (Palo Alto, Calif.: Annual Reviews, Inc., 1976).

    Contains a number of useful, readable articles on coal, coal gasification, geothermal energy, and oil shale.

QUESTIONS

1. Which of the following countries has the largest oil reserves?

    (a) Saudi Arabia   (c) Norway
    (b) United States   (d) Japan

2. Which of the following U.S. energy resources has the greatest potential energy content?

    (a) Oil   (c) Natural gas
    (b) Coal   (d) Tar sands

3. The three largest, major, geothermal power plants in the world are located in the United States, Italy, and Japan. They represent power generated from:

    (a) Hot water.
    (b) Hot water plus steam.
    (c) Hot steam.
    (d) Hot water obtained by pumping cool water down into very deep wells.

4. Can sufficient hydroelectric power be generated from U.S. rivers in the future to produce the amount of energy used in the United States in 1970?

    (a) Depends on an improvement in generator efficiencies.
    (b) Yes, if fuel cells become practical.
    (c) Yes.
    (d) No.

5. The type of geothermal resource that can be least efficiently converted to electricity is:

   (a) Dry steam.
   (b) Wet steam.
   (c) Hot water.
   (d) Hot rocks at a depth of 15,000 ft.

6. The air pollutant that is most dangerous to one's health is:

   (a) Carbon dioxide.   (c) Carbon monoxide.
   (b) Particulate matter.   (d) Oxides of nitrogen.

7. It has been suggested that coal be used to power the automobile of the future. This might reasonably be accomplished by:

   (a) Converting the coal to water vapor.
   (b) Conversion of the coal to power gas.
   (c) Direct burning of the coal in the automobile.
   (d) Making a liquid solution of coal and lime.

8. It is easier to make good estimates of coal reserves than of oil or natural gas because:

   (a) The coal comes in stratified beds or seams.
   (b) The coal reserves are mostly located near the earth's surface.
   (c) The good oil and natural gas reserves are all located more than 5 miles below the earth's surface.
   (d) Most of the nation's coal reserves are located in the Green River area of Utah and Wyoming.

9. What is the moderator in most of the conventional U.S. power reactors?

   (a) Carbon   (c) Heavy water
   (b) Uranium   (d) Ordinary water

10. Which of the following reactor accident scenarios is generally considered to be the most serious?

    (a) One in which a control rod sticks inside the reactor.
    (b) One in which the uranium fuel sticks inside the hollow tube that encloses it.
    (c) One in which a total loss of flow of the normal water supply occurs.
    (d) No serious accident can possible occur.

11. Control of a reactor by means of neutron-absorbing control rods is feasible because:

    (a) A small amount of ^{240}Pu is present in the fuel.
    (b) Not all fission events result in neutrons being emitted.
    (c) Some neutrons from fission events are delayed in their emission.
    (d) Some ^{233}U is mixed in with ^{235}U in the nuclear fuel.

12. The function of the moderator in a reactor is to:
    (a) Provide additional safety.
    (b) Slow down fast neutrons.
    (c) Increase the breeding ratio.
    (d) Provide structural strength.

13. The CANDU reactor uses which of the following as the moderator?
    (a) Heavy water
    (b) Light water
    (c) Carbon
    (d) Natural uranium

14. The emergency coolant system for a nuclear reactor is designed to do which of the following tasks?
    (a) Shut down the reactor in case the emergency safety rods fail.
    (b) Cool off the reactor from the energy generated by the residual fission product activity after a shutdown whereby a loss of coolant has occurred.
    (c) Cool the reactor fuel rods in steady-state operation when the regular water system fails.
    (d) None of the above.

15. What is the fissionable isotope in natural uranium?

16. Most of the U.S. reserves of shale oil rock are located in portions of three states. Name these states.

17. Write the label of the item on the right that is most appropriate for a given topic on the left.

    _____ Light-water reactor fuel
    _____ Heavy-water reactor fuel
    _____ Breeder reactor fuel
    _____ Most abundant isotope of uranium
    _____ Typical conversion efficiency of a coal-fired electric plant
    _____ A common pollutant
    _____ A process that doubles every 12 years
    _____ Basic physics process responsible for energy conversion in nuclear reactors

    (a) Heavy water
    (b) ^{239}Pu
    (c) ^{208}Pu
    (d) 80%
    (e) Carbon
    (f) Particulate matter
    (g) 40%
    (h) Linear growth
    (i) uranium enriched in ^{235}U
    (j) ^{233}Th
    (k) 20%
    (l) ^{238}U
    (m) Exponential growth
    (n) Burning of hydrogen
    (o) Natural uranium
    (p) Water
    (q) Fusion
    (r) Fission

18. Why do the commercial nuclear reactors in the United States, which utilize ordinary (light) water as the moderator, need to use uranium that is enriched in the fissionable isotope?

19. Most of the methods to extract oil from shale rock utilize what other scarce commodity?

20. England and Norway started serious drilling in a promising new area. Where is this new area located?

21. Oil is commonly obtained from wells drilled into the earth. Name three other sources of oil.

22. Name a way to utilize the energy from coal that is essentially nonpolluting?

23. What is the most serious problem associated with the strip mining of coal?

24. Describe what is meant by nuclear fission.

25. What type of geothermal resource is utilized at the Geysers electric plant (located north of San Francisco).

26. What is meant by the term *breeder reactor*?

27. What is the cooling fluid in Canadian nuclear reactors?

28. List two important problems with the production of electricity from our geothermal resource of hot water.

29. It has been said that coal will be used to fuel automobiles, airplanes, and home furnaces of the future. How is this possible?

30. Shale oil is potentially capable of producing all of the U.S. oil needs for perhaps 100 years, but will probably never provide more than a few percent (at most) of our present yearly consumption. Explain.

31. The burning of our limited fossil fuel reserves has been criticized because of the need of future generations to use fossil fuels as raw materials in producing chemicals, synthetic fibers, and so on. Name three energy sources for which this criticism would not apply.

32. One production estimate claims that 3 bbl of water will be needed to make 1 bbl of oil from shale rock. A reasonable value of the surface water available for this purpose in the Green River Basin is 3 million bbl of water per day. Assuming that the U.S. consumption of oil is 6 billion bbl of oil per year, what fraction of this could be taken up by oil from shale rock?

33. According to recent newspaper accounts, the proven natural gas reserves in the Prudhoe Bay Field are about 60 trillion ft^3. In

Chapter 1 we saw that natural gas presently provides about 25% of our annual energy consumption. For how long would the Alaskan Fields cover our total national natural gas needs? (*Hint.* 1 ft^3 of natural gas has an energy content of about 10^6 J.)

34. Would going over to electric cars eliminate the pollution problems associated with this mode of transportation?

35. Suppose that modern pollution control devices on automobiles reduce emissions by a factor of 2 while at the same time gasoline economy is reduced by one-third. What is the real improvement in the amount of auto-related pollutants?

36. What is the principal reason why geothermal resources are best suited for direct-heating use rather than the generation of electrical power?

37. Suppose that we wish to carry out calculations of the energy cost of producing energy for the case of coal. Assume that a unit of coal, which can be burned to produce 100 Btu of heat, can be mined, transported, and so on at an energy cost of 5 Btu of electrical energy. If a coal-burning electric generating plant has an overall efficiency of 30%, what percentage of the output electrical energy is used to produce it?

38. Explain why deep salt bed formations appear to be an attractive location for the storage of radioactive wastes.

39. Which of the following reactions is under serious consideration for a practical fusion reactor on earth?
    (a) $p + p \rightarrow D + \beta^+ + \nu$
    (b) $P + {}^7Li \rightarrow {}^8Be$
    (c) $n + {}^{238}U \rightarrow {}^{239}U$
    (d) $D + T \rightarrow {}^4He + n$

40. Modern fusion reactors will have to operate the plasma containing the fusion material at extremely high temperatures. The reason for this is:
    (a) To get the charged reacting particles to very high velocities so that they can overcome the repulsive electrical force.
    (b) To overcome the retarding influence of the weak force on the fusion reaction.
    (c) To help reduce magnetic field instabilities.
    (d) To permit the breeding of tritium for more fuel.

41. The major technical problem with conventional fusion approaches over the past 15 years has been:
    (a) Production of strong magnetic fields
    (b) Neutron production in the plasma

(c)   Plasma instabilities
(d)   Vacuum problems

42.   Suppose that the fast neutron fission of ^{239}Pu gives rise to 2.0 neutrons per fission event on the average. Could a useful breeder reactor be constructed? Explain.

4 solar history

The past decade has seen a rapid growth in the use of solar energy applications. There is now widespread acceptance of the "solar alternative." This is hardly surprising since solar energy is by far our most plentiful energy source. The technology for putting the sun to work for space and water heating applications is now fully developed. The lack of serious environmental problems has enhanced the image of solar energy.

Except for nuclear power, geothermal energy, and tidal motion, the world's sources of energy are all derived from the sun. This is the case for our fossil fuels, wind power, water power, and various biomass applications. In this book we are investigating the direct use of the sun's energy. Human beings have long realized the tremendous potential associated with the suns rays. It is worthwhile and interesting to look back at our early efforts to capture and utilize the energy coming from the sun.

A. HISTORICAL REVIEW

A newspaper in France in the seventeenth century contained the following account of the destruction of the mighty Roman fleet under the command of Marcellus during the siege of Syracuse in the year 214 BC:

Archimedes set fire to Marcellus' navy by means of a burning glass composed of small square mirrors moving every way upon

83

hinges which when placed in the sun's rays, directed them upon the Roman Fleet so as to reduce it to ashes at the distance of a bowshot.

This account refers to the reputed use of solar energy by the great scientist and mathematician, Archimedes. The veracity of this legend has been disputed by many historians. However, devising such a scheme would have easily been within the powers of Archimedes, who had already distinguished himself by designing a number of innovative war machines, which played a key role in the defense of Syracuse. If we use a rough estimate of the solar energy incident at this latitude, it can be shown that temperatures of 850 to 1000°C could have been attained by an absorbing object located 30 to 60 m away from properly oriented reflecting mirrors of a thousand square meters in area. This concentration of solar energy would have been more than sufficient to ignite the dark-colored sails of Marcellus' fleet in a few seconds. Better documented by historical records was the use of mirrors by Procleus in 626 AD to repel the fleet of Vitellius during the siege of Constantinople.

For the next 1000 years, most of the devices invented to use solar energy were constructed for amusement purposes. An example of this is the solar fountain of de Caus, which is illustrated in Figure 4-1. Several glass lenses were mounted on a frame in order to concentrate the sun's rays on an airtight metal chamber partially filled with water. The sunlight heated the air of this primitive solar engine, forcing water out as in a small fountain.

More serious attempts to make use of the sun's energy soon followed. The French scientist George Buffon constructed the first multiple-mirror solar furnace in 1747. He used an array of 168 flat, square mirrors, each 15 cm on a side, to ignite a woodpile at a distance of 60 m. The famous chemist, Lavoisier, also experimented with solar furnaces because they allowed high temperatures to be obtained in a clean environment. With a special 130-cm-diameter lens filled with alcohol to increase its refractive power, he was able to melt platinum at 1780°C. A schematic diagram of this arrangement is shown in Figure 4-2. The French have maintained an interest in solar furnaces ever since, leading up to the large 1-MW facility of F. Trombe now in use at Odeillo, France. The Italians were not quite so prudent in their work in this field. In 1695, Targioni and Averoni decomposed a diamond by concentrating the sun's light on it with a lens. No more startling discoveries came out of the Italians, and it is assumed that their unwise choice dealt them an unsurmountable financial blow.

Pioneering work with flat-plate collectors was done during the middle of the eighteenth century by the Swiss scientist Nicholas de Saussure. He designed a solar oven consisting of glass plates spaced above a blackened surface enclosed by an insulated box. The sunlight

FIGURE 4-1   Solar water fountain designed by Soloman de Caus in 1615. [Reprinted with permission from *Sunworld*, No. 5 (August 1977), pp. 17–32, published by Pergamon Press, Ltd.]

entered the box through the glass and was absorbed by the black surface; by chemically coating the outside surfaces of the glass, he was able to achieve temperatures as high as 150°C. Because of the low density of the incident solar radiation, this simple approach is still the most inexpensive and reliable way to obtain energy in the form of heat from the sun's rays.

A concentrating type of solar cooker was described in an article in *Scientific American* in 1878 by W. Adams of Bombay, India. As shown

FIGURE 4-2 The burning apparatus of Lavoisier. [Reprinted with permission from *Sunworld*, No. 5 (August 1977), pp. 17–32, published by Pergamon Press, Ltd.]

in Figure 4-3, this apparatus uses a conical, octagonal box lined with silvered glass mirrors to focus light through a cylindrical bell jar onto the food container. This invention worked very well. Another of his suggestions for using solar heat—for the cremation of deceased Hindus and others—was not so well received. A solar cooker that not only employed parabolic mirrors to reach high temperatures but also used a means of heat storage enabling food to be cooked after sundown was developed by Dr. Charles Abbott, the dean of American solar scientists. Dr. Abbott, who died in 1973 at the age of 101, is best known for his early work to develop instruments for analyzing the solar spectrum.

Development of solar-powered engines was begun in the latter part of the nineteenth century by Augustin Mouchot in France and John Ericsson in the United States. Mouchot made a notable advance in collector design by introducing a truncated cone reflector that focused light uniformly along its symmetry axis rather than at a small spot as had previously been done. This device is illustrated in Figure 4-4. In collaboration with Abel Pifre, Mouchot developed a solar steam engine that operated a printing press in Paris in 1882. In the United States John Ericsson, best known for designing the Union Navy's ironclad warship "Monitor" during the Civil War, devoted over 20 years to solar-powered engines, building nine such engines between 1860 and 1883. He invented the Ericsson-cycle, hot-air engine for the conversion of solar heat. As illustrated in Figure 4-5, the sun's rays were concentrated with the use of a parabolic reflector which was designed to track the sun across the sky in order to maintain a constant power output. Ericsson's 1883 model was claimed to develop 746 W for 9.3 m^2 of reflecting surface. He concluded after his last engine that solar-powered engines were 10 times more expensive than conventional designs and therefore not economically justifiable at that time. It has

FIGURE 4-3 The solar cooker of W. Adams. [Reprinted with permission from *Sunworld*, No. 5 (August 1977), pp. 17–32, published by Pergamon Press, Ltd.]

FIGURE 4-4 Mouchot's solar boiler. [Reprinted with permission from *Sunworld*, No. 5 (August 1977), pp. 17–32, published by Pergamon Press, Ltd.]

been reported that by ingeniously converting his solar engines to run on fossil fuels he was able to turn the entire project into a commercial success. Ericsson also measured the energy coming from the sun with an ingenious calorimeter, obtaining a value within 3% of the presently accepted value.

A water pump, energized by a large solar reflector, was built by A. G. Eneas in 1901 to provide water for irrigation on a Pasadena, California ostrich farm. A picture of the Eneas solar pump is shown in Figure 4-6. The solar pump, rated 4 hp, was described in the *Arizona Republican* of February 14, 1901:

The unique feature of the solar motor is that it uses the heat of the sun to produce steam. As "no fuel" is cheaper than any fuel, the savings to be effected by the device is evident.... The reflector resembles somewhat a huge umbrella open and inverted at such an angle to receive the full effect of the sun's rays on 1,788 little mirrors lining its inside surface. The boiler, which is 13 feet 6 inches long is where the handle of the umbrella ought to be. The boiler is the focal point where the reflection of the sun is concentrated. If you reach a long pole up to the boiler, it immediately begins to smoke and in a few minutes is aflame. From the boiler a flexible metallic pipe runs to the engine house near at hand. The reflector is 33-1/2 ft. in diameter at the top and 15 ft. at the bottom.

The entire arrangement was claimed to have cost a mere $2500 and might possibly have been a commercial success had it not been subject to an unfortunate number of mechanical failures, accidents, and eventual destruction as a result of wind storms.

FIGURE 4-5   The tracking solar moter designed by John Ericcson. [Reprinted with permission from *Sunworld*, No. 5 (August 1977), pp. 17–32, published by Pergamon Press, Ltd.]

FIGURE 4-6 Eneas' solar boiler. [Reprinted with permission from *Sunworld*, No. 5 (August 1977), pp. 17–32, published by Pergamon Press, Ltd.]

In 1908, H. E. Willsie and John Boyle, Jr. built a 15-kW solar engine that ran a water pump, compressor, and two circulating pumps in Needles, California. This design, shown in Figure 4-7, was based on their earlier work using a cone-shaped reflector in Olney, Illinois. They used 93 m² of solar collector to heat water, which in turn was used to heat a volatile fluid—in this case sulfur dioxide—which was then vaporized. This device was the first to demonstrate the use of two fluids for solar power generation. Unfortunately, the cost of construction per horsepower was many times that of a conventional steam engine so that the project was a financial failure.

One man, Frank Shuman, came close to making a commercial success of a solar-powered engine. Using a flat-plate collector design of 112 m^2 in area he constructed a successful 3½ hp engine. The solar energy was absorbed by ether, which boiled to provide steam for the engine. Since the flat-plate collector did not have to track the sun, it was a simpler and less costly engine than other contemporary designs. Encouraged by this success, Shuman formed the Sun Power Company and convinced English financiers to back his efforts to build even larger plants. His most noteworthy effort was the construction of a solar power plant near Cairo, Egypt, aided by the technical suggestions of British physicist C. V. Boys. The latter suggested the use of parabolic reflectors placed sufficiently apart so that they wouldn't shadow each other. The solar energy was collected by black painted tubes located along the focal point of the trough-like parabolas, as indicated in Figure 4-8. Because of the concentrating reflector, the collectors were tracked across the sky to follow the sun. This was accomplished through the use of small motors directed by heat sensors and thermostatic controls. The total surface of the collectors was 1200 m^2, and there is authenticated information showing that the output never fell below 39 kW, and averaged around 45 kW. The system fell into disrepair during the World War I, but had it been maintained properly, the irrigational services that it performed would have paid for the cost of construction within three years, after which time the only expense would have been the cost of maintenance.

One of the most practical applications of solar energy involves the production of fresh water. A 4700-m^2 solar still was built in the desert

FIGURE 4-7   The solar-powered plant of Willsie and Boyle. [Reprinted with permission from *Sunworld*, No. 5 (August 1977), pp. 17–32, published by Pergamon Press, Ltd.]

FIGURE 4-8   The solar boiler of Shuman and Boys. This device developed about 45 kW and was used for irrigation purposes near Cairo, Egypt. [Reprinted with permission from *Sunworld*, No. 5 (August 1977), pp. 17–32, published by Pergamon Press, Ltd.]

near Las Salinas, Chile by J. Harding and C. Wilson. It delivered 23,000 liters of fresh water per clear day and operated for about 40 years. More recently, during World War II, solar survival kits were manufactured for emergency use in the South Pacific. The principle of operation of a solar still is demonstrated with the aid of the earth-water design shown in Figure 4-9. The container for the fresh water is placed at the center of the cone-shaped hole. A sheet of clear plastic is stretched over the hole with the edges held down by rocks and earth. A rock is then placed in the middle of the plastic sheet, causing it to slope downward from the sides to the center. The heat generated under the plastic causes the moisture contained in the ground to evaporate, which then condenses on the plastic sheet and runs down into the collector.

As interest in solar energy applications grew, a number of international conferences and symposia were held. In 1953 a symposium on solar energy utilization was held at the University of Wisconsin under the sponsorship of the National Science Foundation. UNESCO and the Indian government sponsored a symposium on solar energy and windpower in 1954. This meeting held in New Dehli, stressed social and political implications as well as solar applications. The International Solar Energy Society was formed in 1954 and in 1955 organized

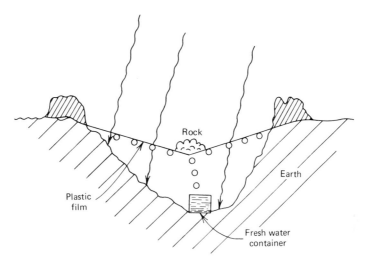

FIGURE 4-9 The energy from the incident solar radiation warms up the space enclosed by the plastic film. This causes water in the ground to evaporate and condense on the thin plastic sheet. (Courtesy of J. I. Yellot, "Solar Radiation and its uses on Earth," published in *Energy Primer*, pp. 4–24, Portola Institute, Menlo Park, Calif.)

two international conferences. The first meeting held in Tucson, Arizona, stressed basic research in the field. This was immediately followed by a symposium held in Phoenix, Arizona. Over 900 people attended the conference and symposium, with 130 coming from outside the United States. In 1961, the Social and Economic Division of the United Nations organized an international conference on solar, wind, and geothermal energy applications. Over 500 scientists attended this conference, which was held in Rome.

These conferences, along with several others held during this period of time, helped to spur interest in solar energy applications. However, the time for solar energy to compete economically with conventional energy sources had not yet arrived. During the meetings in Tucson and Phoenix one prominent official predicted that by 1970 several million homes would be heated using solar energy collection schemes. Unfortunately, these glowing predictions did not materialize—the actual number was more of the order of 10 (not counting houses incorporating passive features).

B. ACTIVE SPACE AND WATER HEATING

Until 1970, the only really successful commercial application of solar energy for space and water heating involved the installation of solar water heaters in Florida and California. Tens of thousands of these heaters were sold in both states until the middle 1950s when the advent of cheap fossil fuel energy caused their sale to decline. It was estimated that more than 50,000 such heaters existed in the city of Miami alone in 1951; but by 1970 the solar water heater had almost vanished in the United States, although it still flourished in Australia, Japan, and Israel. Recent events have now reversed this trend in this country, particularly for swimming pool applications.

In the 1930s interest spread to the topic of house heating. The early experiments at the Zurich Institute of Technology were published by M. Hottinger in 1935. The incident solar radiation was captured by collectors mounted on south-facing roofs or walls with water or air as the heat transfer medium. Insulated tanks of water, rocks, or heat of fusion salts were used for energy storage. Between 1939 and 1960, Professor H. C. Hottel and his colleagues at the Massachusetts Institute of Technology made pioneering studies with a series of four different solar houses. Several other solar houses were built and tested in the period between 1948 and 1960. The most notable were the Boulder, Colorado solar home of G. O. G. Löf, who used air as the heat-exchange fluid, and the Washington, D.C. solar houses of Harry Thomason who invented the trickle-type collector.

Because of the recent rapid growth in the use of flat-plate collector systems (see Chapter 8), one now has available a wide variety of technological approaches from which to choose. However, the basic principles remain the same as those utilized in some of the early solar houses. For this reason it is instructive to examine a few notable solar house examples constructed over the past 40 years. The experience gained in their construction and operation has been invaluable for the present development of the solar industry.

B.1 M.I.T. Solar Houses

In 1938 a Boston businessman, Godfrey L. Cabot, gave $600,000 to the Massachusetts Institute of Technology to be used for solar energy development. Over the next quarter century, M.I.T. constructed four solar houses. The first solar house was constructed in 1939 under the direction of Dr. Hoyt Hottel, a prominent solar pioneer who also played a key role in later designs. The building was a one-story, two-room structure with a floor area of 46.5 m^2. The collector was about 38 m^2 in area, used water as the heat-exchange fluid, had triple glazing, and was tilted at only 30° to the horizontal. A 65,950-liter water storage tank provided annual as well as diurnal storage. It is reported that the tank temperature reached 195°F at the end of the first summer. This solar house was demolished in 1941 to make way for other buildings.

The second solar structure was built in 1947 and was used only briefly for solar experiments. The building had seven south windows. Each one had its own separate collector (vertical orientation) and heat storage system. For example, one combination involved blackened cans located directly behind two cover plates.

This structure was converted in 1948 to the third M.I.T. solar house experiment by enlarging the building and remodeling for occupancy. A 37-m^2 collector, again using water as the heat-exchange fluid, was mounted on the roof at an inclination angle of 57°. In the four coldest months this system provided about 75% of the space heating needs of the 56-m^2 building. Electrical resistance heating was used in the 4550-liter water storage tank as the auxiliary energy source.

The last of the M.I.T. solar houses was constructed in 1959 at Lexington, Massachusetts. The two-story structure, shown in Figure 4-10, had a living space totalling 135 m^2. A portion of the first story was below ground to reduce heat losses. The 59-m^2 solar collector was tilted at 60° to the horizontal and was constructed from 60 panels, each 1.22 m high and 0.81 m wide. The solar thermal collector was a blackened aluminum sheet with two cover plates spaced 1.90 cm apart. Water was circulated through 0.95-cm copper tubes spaced 12.7 cm apart, which were fastened to the absorber plate. Heat storage was provided by a 5690-liter tank. The solar-heated hot water was pumped to a heat exchanger for heating air circulated about the house by a fan. The oil-fired, auxiliary water heater provided a reserve of 57 to 66°C hot water in a 1040-liter tank. Whenever the temperature in the large storage tank was too low to provide adequate heat to the house, the reserve tank was automatically called upon. The house was designed to minimize heat losses, so that when the outside temperature was 0°F, only 32 million J/hr were required. Domestic hot water was also provided by the solar-heated reservoir whenever the temperature was adequate.

FIGURE 4-10 Picture of the fourth M.I.T. solar house constructed at Lexington, Massachusetts. (Courtesy of John Reynolds.)

A thorough analysis of the system performance was made through the two winters of 1959 to 1961. Of the incident solar energy consisting of 311 billion J, about 32% was recovered by the solar thermal collector, and 29% contributed to useful heating of the house. This amounted to about one-half of the space heating needs of the house. Because of maintenance problems this system was abandoned after two years.

B.2 Solaris System

A number of pioneering solar homes were built in the Washington, D.C. area by H. E. Thomason. His first solar house was constructed in 1959 and had 112 m^2 of floor area. Only about 84 m^2 was actually heated in the winter. The solar collector has an area of 70 m^2, part of which was located on a 45° sloping roof, with the other portion located on a 60° sloping south wall. The absorber surface consisted of a blackened, corrugated aluminum sheet located below one layer of glass. About 5 cm of fiberglass was used as the insulating material on the back side. The heat-exchange fluid was water pumped from a 6000-liter storage tank by copper pipe to the top of the collector, where it was then allowed to trickle down the 2.5-cm-wide valleys of the black corrugated aluminum sheet. Maximum water temperatures achieved were in the 52 to 57°C range. For obvious reasons this type of collector is known as a trickle-type system. The water storage tank was a 1.22-m-diameter, 5.2-m-long steel tank surrounded by 50 tons of egg-sized stones for additional thermal storage. The heat-exchange medium for the house was air, heated by circulation around the tank and through the rock-filled chamber. This system provided about 85% of the building's heating needs (according to Thomason).

B.3 Colorado Air System

Dr. George Löf, another prominent solar pioneer, has played a leading role in the development of solar systems that use air as the heat-exchange medium. His first experimental house was constructed in Boulder, Colorado in 1947. This building, with 93 m^2 of floor area, was partially heated by solar energy collected by 43 m^2 of solar panels sloping 27° upward from the horizontal. The solar heat, stored in 8.3 tons of rock, provided about 23% of the space heating needs.

His second solar house project was completed in 1958 and was financed by the American-Saint Gobain Corporation. The house is a nine-room residence comprising 190 m^2 of living space on the main floor. The house, illustrated in Figure 4-11, is of contemporary style. The solar collector consists of two 28-m^2 panels mounted at 45° to the horizontal. Recirculated air passes through the two sections in series, once upward and once downward. Each section of the collector is of the overlapped glass-plate design. Hot air at temperatures up to a maximum of about 79°C passes down to the basement, where it first gives up some heat to the house hot-water supply, and then on either to the bottom of the storage unit or up through the furnace to the distribution system. The heat storage system consists of two cylinders

FIGURE 4-11 View of the solar home of G.O.G. Löf in Denver, Colorado. Air is used for both the collector heat-exchange fluid and the heat delivery system. (Courtesy of John Reynolds.)

of fiberboard each filled with approximately 11 tons of small rocks. The rock column is 5.5 m high, extending from the basement floor to the roof. Almost complete heat transfer from the air to the rocks is obtained from the large surface area of the rock in the storage chambers. The principal problems encountered initially with the design used in the Denver house were excessive glass breakage due to the defective system of glass-plate support for the solar collector and excessive air leakage in the circulating system.

During the winter of 1959 to 1960, a detailed analysis of the daily heating performance was made. In a nine-month period, the total solar radiation on the collector surface was 240 billion J. Of this amount, 171 billion J was of sufficient intensity to warrant collection, and 59 billion J of useful heat was delivered. This represents an overall solar collection efficiency of about 25%. The 59 billion J represented 26% of the total of 280 billion J needed for space and water heating. While this is not a large fraction of the total heating requirement, it should be remembered that the ratio of collector area to floor area was also correspondingly small.

C.   RECENT DEVELOPMENTS

The rapid growth of interest in solar energy applications since 1970 may be partly attributed to concern over dwindling supplies of fossil fuels, but is also due in part to the growing concern over environmental problems. This escalation of interest in solar energy has not been limited just to technical and commercial development, but is also shown by the growing numbers of books, periodicals, and articles being written each year. The International Solar Energy Society, a small group for many years, suddenly found itself deluged with membership applications.

The recent growth of interest in solar applications was matched by a striking increase in federal expenditures in this area during the decade of the 70's. Research and development activities in all types of solar applications benefited from this sudden influx of money. Most of the new solar construction in the first part of the decade used active solar collectors. A modern example of such a system is shown in Figure 4-12.

Most active solar systems also provide preheating of the hot water used in the residence. This allows the solar collector to provide useful work during the entire year. The energy needed to provide hot water can often be 25 to 35% of that required for space heating. Another important contribution that can be made by solar energy is to heat water for swimming pools. Since these are most commonly used during

FIGURE 4-12 The "Decade 80 Solar House" is a prototype solar residence in Tucson, Arizona, built by the Copper Development Association, Inc. The solar energy system provides 100% of the home's space heating and a majority of its air-conditioning and domestic hot water needs.

summer conditions of hot weather, simple, inexpensive solar collector systems can be used. In response to a growing demand for this type of solar heating, a large number of private companies have entered the field. Public utilities are also beginning to recognize the important contribution that solar energy can make. For example, the Public Utilities Department of the city of Santa Clara, California sponsored a program to install solar swimming pool heating systems at a nominal cost. In its utility concept approach, the city owned the solar hardware and charged the customer an installation fee of $200 plus a monthly service fee that averaged about $28 per month from April to September. The monthly rates were guaranteed by the city for a three-year period, after which they could be adjusted to reflect annual cost experience in maintaining and servicing the systems.

As interest and activity in the solar field escalated, a dramatic change occured near the middle of the past decade. People discovered passive solar energy! This passive revolution spread so rapidly that this approach now largely dominates the solar space-heating scene. The dramatic decrease in active space heating systems was confirmed by a 1980 survey of solar installation firms by the Department of Energy. They found that only 7% of the 93,700 active solar systems installed in 1980 were used for space heating. The great majority were utilized for hot water or swimming pool heating.

What is a passive solar building? The answer to this will be given in detail in a later chapter. Here we note only the broad definition of this approach. A passive solar building is one that is designed to use the entire building (floors, walls, windows, and roofs) as the solar collector. A well-designed building functions in harmony with local sun and weather conditions. Spaces and components are arranged so that heat transfer is primarily by natural convection and radiation. The result is an efficient and livable building. Most inhabitants of passive solar residences are impressed by the increased brightness and more comfortable surroundings.

What impact on the nation's energy budget will solar space and water heating have in the future? Clearly the answer to this question depends on whether the present growth rate in the solar industry continues for another 10 to 15 years. If this does occur, then solar energy will make a significant impact by the year 2000, based solely on the growth of the solar space and water heating industry. For solar energy to gain widespread public acceptance, it must be shown to be economically competitive with other forms of energy. To ascertain this the initial cost, plus the small amount needed for maintenance and operation, must be averaged in a suitable manner over the useful lifetime of the system. The present situation for passive solar and solar hot water heating is extremely optimistic. The reasons for this will be detailed in later chapters.

FIGURE 4-13 Light is reflected from the tracking mirrors at the solar power facility near Barstow, California to two separate points. The two points of focus are stand-by points for the heliostat field positioning prior to moving light beams onto (or off of) the solar tower receiver. (Photo courtesy of Southern California Edison.)

If the present growth of the solar industry continues, space and water heating with solar energy can be expected to make a truly significant impact on the nation's energy picture. Based on the fraction of U.S. energy that is annually devoted to space heating and cooling and hot water heating, it is possible that solar energy could contribute as much as 5 to 10% of the total energy needs of the country. But what about the rest? For solar to have a further impact it is necessary that solar electricity be developed.

In 1950, Chapin, Fuller, and Pearson of the Bell Telephone Laboratories developed the first solar cell for the direct conversion of the incident solar radiation into electricity. With the advent of the space age, silicon solar cells rapidly came into prominence as a means to provide power. In 1959, the first successful Vanguard satellite carried 108 solar chips, providing 0.5 W of power. By 1969, 3 million chips providing 105 kW of power were in use in space applications. In the past 10 years considerable progress has been made in overcoming the major problem associated with solar cells, their cost. By 1982 commercial solar cells were readily available at a price of roughly $10 per peak watt. This is still much too expensive for everyday applications, but is in the range where electricity from solar cells is very attractive for remote installations. Recent progress in amorphous silicon solar cells gives rise to continued hope for serious price reductions over the next decade. The possibilities presented by further research and development in the area of photovoltaic devices are so exciting and potentially promising that Chapter 11 is devoted to the subject.

Solar cells are not the only way to generate electricity. The generation of solar electricity through the intermediate step of heating a fluid to a high temperature looks just as promising for the future as direct conversion by means of solar cells. In fact, for large-scale development of solar electricity (plants of the order of 50 MW or more) the solar thermal approach may be better. The rationale and progress in this area is discussed fully in Chapter 10. Progress has now reached the point where a 10-Mw$_e$ (megawatt electric) plant, Solar One, is in operation at Barstow, California. Figure 4-13 shows the beautiful light display that occurs when the plant's reflectors are focused at two separate points away from the central receiver containing the fluid to be heated. When atmospheric conditions are just right (lots of moisture or dust in the air) unusual visual effects can be produced.

REFERENCES

1. F. Daniels, *Direct Use of the Sun's Energy* (New York: Ballantine, 1974).

   This used to be the standard reference for someone desiring an introduction to solar energy. Now available in paperback issue, this useful book contains information on a wide variety of solar topics at an elementary level. Chapter 2 contains a short review of the history of solar energy applications.

2. Ken Butti and John Perlin, *A Golden Thread: 2500 Years of Solar Architecture,* (New York: Van Nostrand-Reinhold, 1980).

   A modern, illustrated history of solar energy developments.

3. J. C. McVeigh, *Sun Power, An Introduction to the Applications of Solar Energy* (New York: Pergamon Press, 1977).

   Contains an informative chapter on historical background.

4. Richard C. Jordan and Benjamin Y. H. Liu, eds, *Applications of Solar Energy for Heating and Cooling of Buildings* (New York: American Society of Heating, Refrigeration and Air Conditioning Engineers, Inc., 1977).

   This excellent publication contains useful chapters on a large number of solar topics. The first chapter, written by the editors, contains some recent historical background.

QUESTIONS

1. Consider the amount of solar energy incident on the roof of a typical house in the Midwest in an average year. It is:
   (a) About one-tenth the average heating needs for the house.
   (b) About equal to the average heating needs for the house.
   (c) About 12 times the average heating needs for the house.
   (d) About 80 times the average heating needs for the house.

2. The Willsie-Boyle solar engine was one of the first to use the two-fluid concept. The second fluid in this engine was:
   (a) Sulfur dioxide.   (c) Methane.
   (b) Carbon dioxide.   (d) Air.

3. The early pioneering development work for the modern flat-plate collector was done by:
   (a) Archimedes.   (c) Nicholas de Saussure.
   (b) G. Buffon.   (d) Charles Abbott.

4. The failure of the commercial solar hot water heating industry in the 1950 to 1970 period can mainly be attributed to:
   (a) Inefficient flat-plate collectors.
   (b) Poor mechanical design.
   (c) Maintenance problems.
   (d) Low cost of natural gas.

5. Willsie and Boyle used what type of solar collector for their solar-powered engine operated at Needles, California?
   (a) A parabolic concentrator.
   (b) A simple flat-plate collector.
   (c) A cylindrical solar collector combined with a mirror concentrator.
   (d) A spherical receiver located at the focus of 168 small flat mirrors.

6. What federal and state incentives favor the purchase of a solar hot water system?

7. In the early twentieth century one man came quite close to developing an economical solar energy application. What is his name?

8. Who pioneered the use of trickle-type collector?

9. Suppose that Archimedes had no mirrors available during the Battle of Syracuse in 212 BC. What readily available piece of equipment could be utilized as a substitute?

10. The number of new housing starts per average year in the United States is about 2.0 million. Assume that exponential growth of the

solar industry will occur, with a doubling time of two years. Starting from 1976, how many years will it take before one-half of the new housing starts incorporate solar heating? About 1000 solar homes were constructed in 1976.

11. The present-day, average cost of energy includes both low- and high-cost contributions. Hydroelectric power is an example of low-cost energy, whereas new coal and nuclear generating plants are examples of high-cost energy. If a solar space heating system is installed, which type of energy (from the cost standpoint) is being saved? Explain how this fact is not taken into account under present economic conditions.

5 the sun

We take the sun for granted. It rises in the morning and sets in the evening without fail. In terms of humankind's residence on earth, the sun is an object that will last forever, continuously radiating away the energy that makes life on our planet possible. The aim of this book is to study in detail how the radiant energy from this, our only renewable energy source, is converted into other useful forms. Before proceeding, it is worthwhile to first answer a few questions about this very important stellar object. When and how was it formed? For how long into the future will the sun continue to radiate at its present rate? Does the sun rotate just like the earth? Does the radiant energy that we receive from the sun come from its interior or from its outermost fringes? What is the magnitude of the solar radiation that we receive? Is the magnitude of this incident solar radiation a constant over long periods of time? These are just a few examples of the many questions that can be asked about our sun.

Our sun is an average star in an average galaxy that we call the Milky Way. The sun moves around the center of this galaxy at a distance of about two-thirds of the maximum radial extent of the galaxy. We are located in a spiral arm of the galaxy in a region that is rich in gas and dust. Here new stars are formed as gravity attracts large masses of this dust into local clusters of matter. Before going on to study the precise nature of our sun let us first take a look at the largest picture, our universe.

A. COSMOLOGY

The enormity of the universe is staggering. It is estimated that the number of visible galaxies in the universe is in the billions; there are so many that the number cannot be directly counted. The sun is only one of the 10^{11} stars in the Milky Way. The diameter of our galaxy is about 100,000 light-years. A light ray originating on one side of our galaxy would take 100,000 years to cross to the other side. Within a distance of 2×10^6 light-years are located another 15 galaxies in our own local cluster of galaxies. An example of the enormity of space and local clustering is shown in Figure 5-1.

How was the universe formed? A number of hypotheses have been advanced. One of these states that the universe was created almost exactly as we now observe it. However, between 1920 and 1930, astronomical evidence that was in sharp disagreement with this model began to accumulate. It was observed that everything in the universe appears to be moving away from us. The speed of recession of an object appears to increase as the distance away from us becomes larger. Evidence for this comes from observations of the Doppler shifting of the light received from these distant objects. This phenomenon is a shifting of the color of the observed light toward the red end of the spectrum because of the motion of the source of radiation away from us. It is a very well-understood physical effect and occurs for all types of wave phenomena (sound, light, radio waves, etc.). It is as though we were at a point on the surface of an expanding balloon, with every other point receding from us as the balloon expanded. This experimental observation of an expanding universe is in sharp disagreement with the steady-state hypothesis.

About 30 years ago George Gamow proposed that the universe originated by having all its mass concentrated in a relatively small volume. The concentration of matter was so high that a temperature greater than 10^{16} K was achieved. The radiation pressure from this primordial fireball was tremendous and it exploded outward. Thus, our universe was created in a "Big Bang"! This event is supposed to have occurred over 10 billion years ago and provides a ready explanation for the theory of an expanding universe. The parts of the fireball with the greatest relative velocities are now concentrated in the distant galaxies that we see receding so rapidly from us.

The Big Bang model does not imply that there exists a center to the universe. In fact, there is no point in three-dimensional space from which all matter issued. The Big Bang marks the onset of an expansion of space and time itself, an expansion that is hard to visualize since we cannot imagine a fourth dimension. The distance between any two points increases just as the distance between spots painted on a balloon's surface will separate as the balloon is inflated.

FIGURE 5-1   Photograph of the large cluster of galaxies in the constellation Hercules. A cluster may contain anywhere from two to several thousand galaxies. The average distance between galaxies is about 1 million light years. (Courtesy of Hale Observations, Catalog #68).

This suggestion was not widely accepted until recently when careful measurements with very sensitive detection equipment measured what is now believed to be the background radiation left over from the original big bang. The intensity of this background radiation varies at different wavelengths and was found to have a distribution almost precisely that to be expected if the major portion of the universe were a blackbody emitter (see Chapter 6) at a temperature of about 3.0 K. The same distribution of intensity is obtained no matter in which direction the special radiofrequency detectors are pointed. According to theory, this type of distribution is exactly what would be expected to be left over from a Big Bang occurring so long ago. As the universe expanded, the density of matter decreased and the equivalent temperature of the universe fell. When conditions were cool enough, protons and electrons precipitated into hydrogen atoms, allowing the radiant energy left over to pass about without being immediately absorbed. This cosmic background radiation, detected at microwave frequencies, constitutes the most important evidence for the Big Bang hypothesis.

How are stars formed from this expanding cosmic dust? The answer to this is still not certain, but it is believed that stars condense from this interstellar material under the influence of gravity. Suppose there is a large cloud of gas, mostly in the form of molecular hydrogen. Occasionally some kind of shock or gravitational wave triggers the gas in a particular region of interstellar space to start forming large masses of this dust. Picture one large mass beginning to collapse. The material at the center increases in density faster than at the edge. There is a core of rapidly contracting material surrounded by a slowly collapsing envelope. Gravitational energy heats the core until, at its center, temperatures reach the ignition point of fusion reactions. The more slowly contracting envelope blocks out the birth of the star from our view. Dust in the envelope cuts out the light from this new star. However, the radiation absorbed by the dust heats it to the point that infrared radiation is emitted. Many such protostars have recently been observed lying embedded in large interstellar clouds of molecular hydrogen.

The life history of a typical star is complicated and can be illustrated with the aid of the diagram shown in Figure 5-2. This is a plot of the relative brightness of a star as a function of its surface temperature. Most stars are found along the main sequence, which is shown as a dashed line along the figure. The main sequence has the most massive stars on the upper left and the least massive on the lower right. The more massive the star, the greater its temperature and luminosity. Initially the protostar lies to the right of the main sequence. As the new star shrinks in size, its luminosity increases. When the protostar gets most of its energy from thermonuclear reactions (see next section), rather than gravitational contraction, it reaches the main sequence. The total time from the initial collapse to arrival on the main sequence is about 50 million years.

The average star with a mass about equal to that of the sun spends roughly 10 billion years on the main sequence line of the diagram. At the end of this time it has used up almost all of the hydrogen fuel in the core, or about 25% of the total mass of the sun. As the core contracts further, it heats up the surrounding hydrogen which now burns at a very fast rate. The radius of the star increases more than a hundredfold

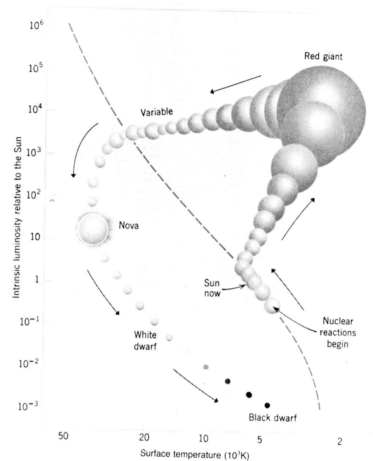

FIGURE 5-2 The life cycle of a typical star such as our sun is depicted schematically upon this Herzsprung-Russell plot. It will be another 5 billion years before the sun enters the red giant stage. (Courtesy of John Wiley & Sons, *Physics and the Physical Universe*, by J. Marion.)

and the star attains red giant status. This giant star has a surface temperature somewhat below that of the sun and the light emitted peaks in the red portion of the spectrum.

In our solar system the starting of the red giant period will occur in about 5 billion years. It would then take an average star like our sun only a few hundred million years more to attain full red giant status. At this time the core temperature will have reached the minimum necessary for the helium in the core to undergo fusion reactions. This helium burning occurs quickly and lasts only a few thousand years. This whole process is very unstable and occasionally the star undergoes a violent eruption with the cool outer layers thrown out into space. This is commonly called the nova phase. The interior region continues burning its nuclear fuel into exhaustion, finally reaching the status of a "white dwarf" star.

The most massive stars have life histories that are quite different from those of the average stars. They shine brilliantly for a few million years and then die very spectacularly. Rather than having many relatively gentle ejections of material as do the smaller stars that undergo the nova phase, these giant stars often finish their life history very quickly in one giant explosion called a *supernova*. The light intensity from this event can attain a brightness 10,000 times greater than that of a typical nova. In the past 2000 years seven of these huge stellar explosions in our galaxy have been authenticated from a close scrutiny of historical records. From observations of supernovas in other galaxies, it is known that such catastrophic events are rather frequent— about one per century in a galaxy such as ours. Because this type of explosion tends to occur near the central plane of a galaxy, and since the central plane of our own galaxy is filled with obscuring dust, very few supernovas have been observed over the span of recorded history. Supernovas play an important role in the formation of the heavy elements. The heavy elements (formed by neutron-capture reactions) are ejected into space during the explosion. At some later time this heavy material is incorporated into the next generation of stars.

Our universe contains energy in many different forms such as heat, light, gravitation, and nuclear energy. Stored chemical energy, with which we are so familiar because of our large fossil fuel reserves, counts very little in the universe as a whole. The dominant form of energy is gravitational energy. Every bit of matter attracts all other matter through the force of gravity. The gravitational energy, which separated masses possess, can be converted into light and heat by allowing the originally separated masses to fall together. Because of the large amount of mass in the universe, gravitational energy greatly outweighs all other forms.

When two masses are separated the laws of physics favor their coming together under the influence of the gravitational force. If this is the favored situation it is then logical to ask why gravitational energy is still predominant after 10 billion years? Why has the universe not collapsed under gravitational contraction?

The answer to this question is difficult to give in a simple manner; essentially one can say that the universe survives because of a number of constraints (see reference 3 at the end of this chapter). At the galactic level the important constraint is the large size of the universe. If a finite volume is filled with matter to a density ρ, then the matter inside the volume cannot collapse in a time shorter than the time, T, it would take a mass, m, to "fall" across the given volume under the influence of the attractive gravitational force of the rest of the mass. Using a density, ρ, of 1 atom/m^3 as an average value for the entire universe, a value of $T \cong$

100 billion years is obtained. This is a factor of 5–10 longer than the estimated age of the universe. It appears that on a global scale the universe is safe from gravitational collapse. At the level of individual galaxies the situation is considerably different. The density inside our galaxy is a million times that of the universe as a whole, resulting in a free-fall time of only 100 million years. Since our galaxy has existed much longer than this, it is clear that some other constraint must be influencing the situation. This is the "spin" constraint, which results from the orbital motion of the stars about the center of the galaxy, illustrated by the photograph in Figure 5-3. In a similar way it is the orbital motion of the earth and the other planets that keeps them from falling into the sun.

FIGURE 5-3 The constraint due to orbital motion of the stars in a galaxy is illustrated in this photograph of M104 taken with the 500-cm telescope on Mt. Palomar. The circular paths of the individual stars about the galaxy center keep it from collapsing. (Courtesy of Hale Observatories, Catalog #32).

B. SOLAR FUSION REACTIONS

As a typical star, the sun was formed by the mutual gravitational attraction of its constituents. When the individual particles condensed from the cosmic dust cloud and "fell" together, their kinetic energy increased, exactly as a falling stone picks up speed. As the size of the sun decreased, gravitational energy was converted to energy of motion, heating up the interior of the sun to very high temperatures, of the order of 2×10^7 K. What keeps the sun from collapsing even further is that at these high temperatures thermonuclear reactions involving the burning of hydrogen can take place, and these reactions release large amounts of radiation. The pressure from this radiation is sufficient to keep the sun in equilibrium at its present dimensions and to prevent further collapse. About 5 billion years in the future, when the sun's hydrogen is spent and this pressure is removed, the sun will again gravitationally contract. At that time, the resulting higher temperatures will permit the burning of new fuels, helium and carbon, as the sun enters into red giant stage.

What are the nuclear reactions taking place in the fiery interior of the sun? Since the sun is about 90% hydrogen you should not be surprised that the basic process involves the fusion of hydrogen to form stable helium, with the accompanying release of a large amount of energy. This basic process, first postulated by physicist Hans Bethe, is called the proton-proton cycle. The conversion of hydrogen to helium actually involves three separate reactions. The first reaction is of particular importance to us because it consists of the interaction of two hydrogen nuclei to form deuterium. Along with the deuterium, a positive electron and a massless particle called a neutrino are also emitted. This process can be written schematically as

$$P + P \rightarrow D + \beta^+ + \nu \qquad (5\text{-}1)$$

The deuterium formed will immediately react with the hydrogen in the sun, through a nuclear interaction. The result of this process is the formation of the light isotope of helium, ^{3}He, and an energetic radiation called a gamma ray. As the ^{3}He is formed its density finally becomes large enough that occasionally two such atoms will collide and interact to form ordinary helium, ^{4}He, and two nuclei of hydrogen. These two reaction processes can be summarized as follows:

$$D + P \rightarrow {}^3\text{He} + \gamma \qquad (5\text{-}2)$$

$$^3\text{He} + {}^3\text{He} \rightarrow {}^4\text{He} + 2P \qquad (5\text{-}3)$$

The entire process can be summarized by saying that four hydrogen nuclei combine to form helium through the processes described by the above three reactions. This fusion process releases the energy needed to power the sun.

Why doesn't this process, starting with the reaction of Eq. 5-1, take place catastrophically with the entire sun exploding in one giant blast? The answer to this question lies in another constraint of nature. We have previously discussed examples of three of the four fundamental forces in nature: the gravitational, electrical, and nuclear forces. It is the fourth basic force, called the weak interaction, that is responsible for the reaction, described by Eq. 5-1, taking place. This weak interaction is responsible for any reaction that occurs where an electron or a neutrino is emitted. Because the weak interaction is only roughly 1 million-millionth as strong as the nuclear force, this process proceeds at about a factor 10^{-12} as fast as most nuclear reactions. This is so slow that the only way the reaction described by Eq. 5-1 can play a significant role is for a very large amount of hydrogen to be available—exactly the situation that prevails in the sun. Because of the slowness with which the above reaction proceeds, the formation of deuterium from proton-proton capture has never been observed in the laboratory.

The large amount of hydrogen that is necessary to have an appreciable energy release by this process precludes using this reaction for power on earth. The present attempts to make a controlled fusion reactor are based on the fusion of the deuteron, D, with atoms of tritium, T, an isotope with the same electrical charge as hydrogen but with three times its mass.

C. PHYSICAL DESCRIPTION OF THE SUN

The sun is 1.39 million km in diameter and is at an average distance of 150 million km from the earth. The sun is a peculiar hydrodynamic object, with the equator rotating about its axis in 26 days, while the polar regions rotate about this axis once every 37 days. The temperature of the hot interior is estimated to be near 2×10^7 K, which is high enough that the fusion of hydrogen to helium can take place. This hot, interior region where most of the heat generation takes place is thought to extend out to $0.23R$, where R is the radius of the sun. It is estimated that this core region where most of the sun's fusion reactions occur contains about 40% of the mass of the sun and provides over 90% of the total heat generation.

The energy generated in the interior is transferred slowly out to the surface through a succession of radiative processes in which emission, absorption, and reradiation occur. About 1 million years is required for this energy transport to take place. The radiation that finally reaches the earth comes from a narrow, cooler surface region called the photosphere. This is a region of low-density (about 10^{-4} that of air at sea level) ionized gases that is rather opaque to visible light. Outside the photosphere are found three distinct, almost transparent regions. The first region, called the reversing layer, is a shell that is several hundred miles deep containing much cooler gases. Next is found a much thicker, less dense layer about 6000 miles in depth called the chromosphere with a temperature somewhat above that of the photosphere. The last layer is the corona, a region of very low density and very high temperature, of the order of 2 million K. The various regions of the sun are shown schematically in Figure 5-4.

The radiation from the sun, while approximating that from a blackbody (see Chapter 6), differs slightly primarily because of the structure in the surface-emitting region and the absorption in gases at the surface. For most purposes we can consider the sun as a perfect radiator, emitting at a temperature of about 5800 K ($\cong$ 11,000°F). The wavelength range of 300 to 4000 $\times$ 10^{-9} m (10^{-9} meter is called a nanometer) includes most of the energy of the solar radiation. The spectral distribution of the energy emitted by the sun is of particular importance to the useful collection of solar radiation because of the dependence of the reflective and absorptive properties of materials such as glass and various metals on the wavelength of the radiation.

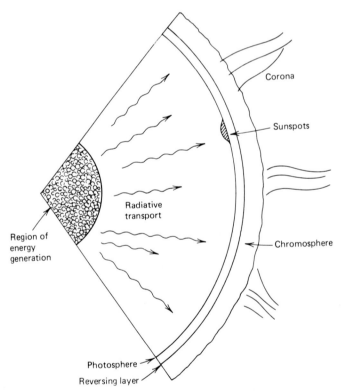

FIGURE 5-4 Schematic illustration of the various regions of the sun. As explained in the text, most of the energy generation from the fusion of hydrogen occurs in the central core of the sun.

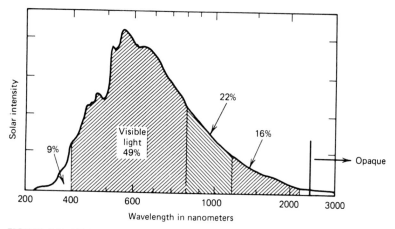

FIGURE 5-5   Division of the intensity of the solar spectrum into its different wavelength regions. About 49% of the solar energy is in the region of visible light, 9% is in the ultraviolet region, and about 4% has a wavelength long enough that it cannot pass through a thin layer of glass. The line of 2500 nanometers indicates the point where ordinary glass becomes opaque.

The distribution of the amount of solar energy incident on the earth as a function of wavelength is shown in Figure 5-5. About 80% of the total incident solar radiation lies in the wavelength region which is transmitted by ordinary glass.

D.   SOLAR PUZZLES

In the past 10 to 15 years numerous new developments have occurred in astronomy because of improvements in optical observational techniques and the development of radio and infrared astronomy. Such fascinating objects as quasars, pulsars, and radio galaxies have been found. The sun has also been scrutinized in a variety of new ways. One result of all of this new, accumulated data is the indication that things within the sun are not as well understood as was previously thought. For example, the explanation of the protruding structure shown in Figures 5-6 and 5-7 is most likely that plasma is trapped in its strong magnetic and gravitational fields.

The first step in solving this and some of the other mysteries of the sun is to obtain a basic understanding of its magnetic properties. It has long been known that the sunspots on the surface of the sun follow an 11-year cycle. The exact mechanism responsible for this cyclic behavior is unknown, but in recent years considerable progress has been made. It is now believed that at the beginning of its cycle the sun has a weak magnetic field generated by flowing streams of charged particles.

FIGURE 5-6 Photograph of spectacular prominence taken at California Institute of Technology Big Bear Observatory. Taken from the light of hydrogen, it shows the visible surface of the sun—the photosphere—together with the lower atmosphere—the chromosphere. The pattern of markings on the surface represent a cell structure that is 32,000 km in diameter. (From *Aviation Week and Space Technology*, January 14, 1974.)

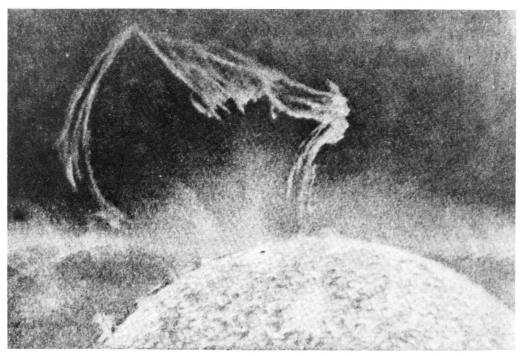

FIGURE 5-7 Puzzling solar eruption sends huge mass of helium surging up 800,000 km from the surface of the sun. Scientists are presently unable to explain this phenomenon, where a cloud of cool gas seems to run into an invisible wall in hot corona. The influence of magnetic field and gravity alone cannot account for the original explosion, stonewall effect and suspended "rain." (From *Aviation Week and Space Technology*, January 14, 1974.)

Because the sun is by no means a rigid body, some areas on its surface and in the interior may rotate at different speeds. This differential rotation twists the magnetic fields into a tangled pattern beneath the surface. At times the magnetic field pops through the surface and causes sunspots.

Sometimes when the magnetic field lines stretch from one sunspot to another, hot plasma can get trapped inside. In this way a tremendous amount of energy can be stored. For example, the trapped energy can be as large as the United States would use, at present rates of consumption of energy resources, in 100,000 years. When the stored energy reaches a critical level, a solar flare, such as that illustrated in Figures 5-6 and 5-7, can occur. Among the many questions still unanswered is how so much energy can be gathered and stored.

Even the shape of the sun is now in question. As shown in Figure 5-8, the sun appears to be a perfect sphere. However, careful measurements by Dicke and Goldenburg of Princeton in 1967 indicated that the sun is slightly oblate, that is, fatter at the equator. This deviation from sphericity, if true, could have significant implications for the detailed

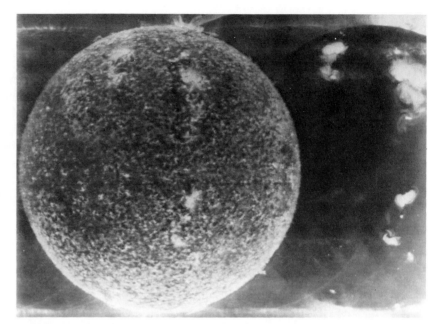

FIGURE 5-8   The solar disk appears to be a perfectly spherical object. The possibility of a slight deviation from this symmetrical shape could have important consequences for general relativity. (From *Aviation Week and Space Technology*, January 14, 1974.)

form of general relativity. Later experiments are at variance with the Dicke result, however, so that the entire matter is an open question at present.

Perhaps the most puzzling aspect about the sun is related to our present understanding of its basic nuclear processes and hydrodynamics. Coupling these two aspects of physics together has led to an understanding of the energy generation and radiative flow inside the sun that appeared at one time to be in very satisfactory agreement with all measured properties. However, one cannot directly study the interior of the sun. The light energy generated in the core of the sun reaches the surface only after a large number of processes have occurred. For example, the light energy can collide with atoms and be scattered. Or, it may be absorbed and reemitted by the matter that makes up the bulk of the sun. These processes occur many, many times so that the light coming from the surface of the sun corresponds to radiant energy that left the center more than a million years ago. This means that when we look at the sun, the light emitted tells us only about the solar surface and reveals nothing about the solar interior.

If the proton-proton process described earlier is the dominant fusion cycle in the sun, then occasionally an isotope of Boron, ^8B, will be formed. ^8B is not stable and will immediately undergo radioactive decay; one of the decay products is a massless particle that is called a neutrino. Unlike light, the neutrino has no appreciable interaction with matter. When a neutrino is emitted in the interior of the sun it reaches the surface 2.3 seconds later. Since it only takes 8 minutes and 20 seconds for the neutrino to cross the 150 million km between the earth and the sun, it is possible to study the interior of the sun if somehow we can only detect the neutrinos. Of course, since the neutrino interacts so weakly with matter, it is very difficult to detect. Fortunately, it turns out that ^8B neutrinos have a relatively high probability (for neutrinos) of interacting with chlorine (Cl) atoms. When a neutrino interacts with a chlorine atom, an electron is expelled and an atom of argon (A) is formed. The argon atom is radioactive and easy to detect. Over a decade ago, the physicist R. Davis of Brookhaven National Laboratory initiated an experiment to measure the number of solar neutrinos striking the earth. Davis' apparatus is located 1440 m below ground level in the Homestake mine in South Dakota. His detector consists of 378,000 liters of carbon tetrachloride (ordinary cleaning fluid). Many years of careful measurements have led to the conclusion that the number of neutrinos detected is only one-third that expected on the basis of present solar models. The uncertainties in both the calculated and measured rates of neutrino events are far too small to allow any possibility of agreement between the two values.

Does this mean that we are in error concerning our present understanding of the reaction processes in the sun's interior? Or is there another explanation? A number of hypotheses have been suggested. One highly intriguing suggestion is that the neutrinos decay on the way from the sun into another neutrino form which cannot interact with the ^{37}Cl isotopes of the cleaning fluid. This possibility is of great importance also to the present understanding of elementary particle physics.

Another suggestion is that the sun may undergo a period of reduced power generation. It is estimated that a 1% drop in the radiative output from the sun would lower the world's average temperature by 1.1°C. This would have a striking effect on the world's climate; a fall of one-half degree per century for a few hundred years would be enough to send the world into an ice age. In fact, it is now generally believed that the various ice ages occur at intervals varying from 20,000 to 100,000 years, and that the responsible mechanism is the variation in summer sunshine because of cyclical changes, both in the orientation of the earth's axis in space and in the shape of its orbit.

Recent modern experiments to study the variation in the sun's luminosity have been made with a high-precision radiation sensor located on board a satellite. This experiment, directed by scientists from the Jet Propulsion Laboratory, is capable of detecting fluctuations in the solar output as small as 0.001%. Measurements taken over a four-month period show fluctuations in the sun's intensity at the level of 0.05%, except for two separate occasions when it jumped by 0.2%. On both occasions rapid growth of sunspot groups was observed.

The possibility of fluctuations in the sun's output has triggered many investigations of past historical records related to such things as sunspot activity and solar output. It is very intriguing that in 1893, E. Walter Maunder, superintendent of the Royal Greenwich Observatory in London, after a search through old books and journals, postulated that the sun was not the regular and predictable star that had previously been believed. Maunder found that between 1645 and 1715, sunspots and other solar activity had all but vanished from the sun; the total number of sunspots recorded in this 70-year time interval was less than what is seen in a single average year today. Recent studies of this time period have vindicated Maunder's findings. The time period of 1645 to 1715 was also one of practically no aurora borealis (northern lights) activity and no coronal streamers.

There is a very powerful, modern technique that sheds further light upon this subject and extends the discussion to very ancient times. Radioactive ^{14}C is formed in the upper atmosphere through the action of cosmic rays. When the sun is very active its extended magnetic field shield the earth from cosmic rays, with the result that little ^{14}C is formed. When the sun is less active, as during the Maunder minimum, the reverse is true. It is fortunate that trees provide a record of the amount of ^{14}C in the atmosphere. By analyzing the wood in the annual growth rings of very old trees, the ratio of ^{14}C to the common isotope of carbon, ^{12}C, can be determined. It has been found that the time period between 1645 and 1715 was a period of greatly enhanced ^{14}C intake by trees. From an analysis of the ^{14}C data over the past 5000 years it is found that there have been at least 12 solar excursions as prominent as the Maunder minimum. Another important aspect of the Maunder minimum time period is that it corresponds almost exactly with the coldest part of the "little ice age," a period of unusual cold in Europe from the sixteenth century to the early nineteenth century. During the Maunder minimum, Alpine glaciers advanced farther than at any time since the last major glacial period 15,000 years earlier. A search of old records of Alpine glacial advances and retreats correlates exactly with periods of greater or lesser sunspot activity as determined from the ^{14}C data. Therefore, it appears that for the past 5000 years, all climatological curves rise and fall in response to the long-term level of solar activity.

REFERENCES

1. M. Zeilik, *Astronomy: The Evolving Universe*, (Harper & Row, 1979).

    Contains much information relevant to the topics covered in this chapter. Material is written in an informative, enthusiastic, and nonmathematical manner.

2. J. B. Marion, *Physics and the Physical Universe* (New York: Wiley, 1971).

    Chapter 17 of this elementary physics text provides an introduction to the subject of astrophysics and cosmology. Material covered includes the basic stellar nuclear reactions, formation of the elements, nonoptical astronomy, stellar evolution, and the big-bang hypothesis.

3. F. J. Dyson, "Energy in the Universe," *Scientific American*, 51 (September 1971).

This informative article on energy in the universe appears in the 1971 Energy issue of *Scientific American*.

4. N. Robinson, *Solar Radiation* (Amsterdam/London/New York: Elsevier, 1966).

The first chapter of this book presents a good summary of the physical properties of the sun.

QUESTIONS

1. It is now believed that the big-bang hypothesis about the origin of the universe is the correct one. The best evidence for this hypothesis is:

(a) The presence of an isotropic, blackbody radiation of temperature that is 3.0 K.

(b) The presence of great amounts of cosmic dust in our galaxy.

(c) The presence of very high-energy cosmic rays striking the earth.

(d) The sighting of quasi-stellar objects (quasars) in the last few years.

2. Most of the energy of the universe results from:

(a) Nuclear reactions in stars.

(b) Supernova events.

(c) Gravity.

(d) Nuclear fission.

3. Our galaxy (Milky Way) would ordinarily collapse under the force of gravity if it were not for:

(a) Radiation pressure from the center.

(b) Rotational motion of all the stars about the galactic center.

(c) Centrifugal force exerted by nearby galaxies.

(d) Some unknown force.

4. The sun exists in a stable equilibrium between the force of gravitation and:

(a) The centrifugal force due to its rotation.

(b) The outward force due to nuclear fission.

(c) The outward radiation pressure from fusion reactions.

(d) The weak interaction constraints.

5. One of the following is not a useful fusion reaction for an earth-based fusion reactor:

(a) $D + D \rightarrow {}^3He + n$

(b) $D + D \rightarrow T + P$

(c) $P + P \rightarrow D + \beta^+ + \nu$
(d) $D + T \rightarrow {}^4He + n$

6. The two major constituents of the sun are hydrogen and helium. The percentage of helium (by mass) in the sun is about:

(a) 80%.
(b) 90%.
(c) 20%.
(d) Unknown at present.

7. The "red shift" of light received from distant stars is:

(a) Due to their gravitational collapse.
(b) Evidence that the universe is expanding.
(c) Early evidence that was thought to be proof of a steady-state universe.
(d) Due to cosmic background radiation.

8. The direct radiation that we receive from the sun has its origin in the:

(a) Photosphere.
(b) Corona.
(c) Chromosphere.
(d) Sun's interior.

9. When the sun enters the red giant stage, the following will be true:

(a) The sun's interior temperature will go to 5×10^6 K.
(b) Almost all of the hydrogen in the core will have been consumed.
(c) The radius of the sun will be greater than the orbital distance to the planet Neptune.
(d) The sun's major constituent will be carbon.

10. Most of the sun's energy results from:

(a) Nuclear reactions.
(b) Gravity.
(c) Chemical reactions.
(d) Fission processes.

11. In the red-giant stage a star dominantly burns which of the following?

(a) Hydrogen
(b) Helium
(c) Carbon
(d) Silicon

12. The energy spectrum from the sun resembles closely that of a perfect emitter radiating at a temperature of:

(a) 5800 K
(b) 11,000°C
(c) 59,000°C
(d) 590°C

13. The sun is located out near the rim of the Milky Way. Why does it not fall in toward the center of our galaxy?

(a) Because of its orbital velocity about the galactic center.
(b) Because the force of the gravitational attraction with the rest of the stars in the galaxy is too weak for collapse.

(c) The radiant pressure due to radiation from the center of the galaxy is larger than the attractive gravitational force.
(d) Because not enough time has passed for this to occur.

14. If we could observe light from a star located on the opposite side of the Milky Way, how long ago would this light have been emitted from the star?

(a) 100 years
(b) 80,000 years
(c) 10 million years
(d) 4.6 billion years

15. In this problem you are to match up the basic forces on the left with the examples on the right. Put the letter of the example you choose in front of the force.

FORCES	EXAMPLES
—— 1. Nuclear	(a) A falling weight
—— 2. Electromagnetic	(b) Emission of light
—— 3. Weak	(c) $P + P \rightarrow D + \beta^+ + \nu$
—— 4. Gravitational	(d) $D + T \rightarrow {}^4He + n$

16. The following short questions concern the physical nature of our sun.

(a) For how many more years should the sun continue to burn hydrogen in its core at the present rate?
(b) What common element will the sun burn after the hydrogen in its core is used up?
(c) In case (b) above will the radius of the sun be greater than, equal to, or less than its present radius?
(d) Is the period of rotation at the equator the same as that near the poles, or is it different?
(e) What common effect is associated with the penetration of the sun's interior magnetic field through the surface?
(f) Approximately how long does it take for radiative energy generated in the interior of the sun to be transported to the surface.
(g) The sun behaves like a blackbody emitting at what temperature?
(h) Observers on earth have traced out a periodic cycling of the sunspot activity. What is the period of time between the point of maximum activity of one cycle and the maximum for the next cycle?
(i) If the sun were not perfectly round it would have serious implications for what important theory of nature?

17. Why is it that the basic fusion process in the stars will not work on earth?

18. Explain why the sun does not collapse into the center of our galaxy under the influence of gravity.

19. What is the mass of the neutrino?

20. Put a cross in front of each of the following reactions which occur because of the presence of the weak interaction in nature.
(a) $P + P \rightarrow D + \beta^+ + \nu$
(b) $D + P \rightarrow {}^3He + \gamma$
(c) ${}^3He + {}^3He \rightarrow {}^4He + P + P + \gamma$
(d) $n \rightarrow P + \beta^- + \bar{\nu}$
(e) ${}^4He + {}^4He \rightarrow {}^8Be$
(f) ${}^3H \rightarrow {}^3He + \beta^- + \bar{\nu}$

21. All attempts to explain the inability to detect solar neutrinos have so far been unsuccessful. Nevertheless, a large number of reasonable hypotheses have been put forth. Although quantitative calculations must be left to experts in the field, we can speculate qualitatively about possible solutions. Give at least three plausible resolutions to the neutrino problem.

22. An interesting development in modern astronomy has been the discovery of *quasars*. These bright, starlike objects have such large red shifts that astronomers generally believe they are located at the very extreme outer limits of the visible universe. At these large distances, their light output would have to exceed our entire galaxy of 10^{11} stars. Suggest a possible alternative not requiring such a massive energy usage to account for the observed light intensity.

23. An article published in June 1976 in *Scientific American* discusses the seven possible supernova events in our galaxy over the past 2000 years. Read this article and tabulate the evidence in favor of this classification for each possible event. Would any of these criteria apply to supernova events outside of our galaxy.

24. Suppose that the interior of the sun out to $0.6R$ were rotating roughly six times as fast as the sun's surface. In this case how much faster is a particle moving at a distance of $0.5R$ than one at the surface?

25. The sun and the planets of our solar system are thought to have all originated from the condensation of galactic dust about 4.6 billion years ago. Can you think of any reason why the sun is composed mainly of hydrogen, whereas the earth is now composed mainly of heavier elements, such as iron?

6 solar radiation

Any discussion of the practical applications of solar energy must begin with an understanding of the basic resource, solar radiation. To do this it is necessary to first study the basic physical principals governing the generation and propagation of this energy from our sun. After this effort, the steps necessary to capture solar radiation and turn it into a useful energy resource can be outlined and more easily understood.

The field of solar radiation is deeply indebted to Charles Greeley Abbot (1872–1973). He dedicated a major portion of his life to the measurement of solar constant (the luminosity of the sun at the top of the earth's atmosphere). As director of the Smithsonian Astrophysical Observatory, his efforts were largely responsible for the perserverance and success of the solar constant study. Most of these measurements were made at a remote observing station at Mt. Montezuma in the Atacama Desert in northern Chile, because of the clear skies at this high-altitude site. It was Abbot's dream to predict climate changes from knowledge of the supposed variability in the sun's output. Although this practical goal was not met, the standard for precise solar radiation measurements was set.

Since solar radiation is one example of a general phenomenon known as wave motion, we begin with a discussion of waves. This is followed by a brief review of some of the important properties of light, the wave motion of crucial importance to us. Then the energy emitted by hot objects is considered. The light emitted by the sun follows closely the emission of radiation by a "perfect" emitter at a temperature of 5800 K. It is important to understand exactly what constitutes a perfect emitter and to know the laws that govern the radiation given off. The rest of the chapter is devoted to studying the practical aspects of measuring and quantifying the solar intensity coming from the sun.

A. WAVE MOTION

Consider what happens when a pebble is dropped into a quiet pond. A circular wave pattern spreads out from the point of impact. An important property of this wave is that the water does not move forward along the direction of the wave motion. A piece of wood placed on the water moves up and down as the wave passes as shown in Figure 6-1 and does not travel along with the wave. This type of wave motion is called a transverse wave because the medium through which the wave passes vibrates in a direction perpendicular (transverse) to the direction of motion of the wave. It is an important property of wave motion that the medium does not move along with the wave. Instead, it is the disturbance that we term a wave that propagates.

There is another important type of wave motion. This is longitudinal wave motion, the common example being sound. A vibrating diaphragm such as a loudspeaker produces a wave disturbance that travels outward. While this wave motion moves outward, the air molecules

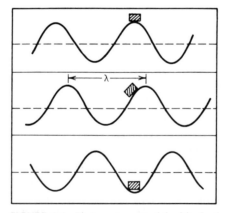

FIGURE 6-1 The movement of the block of wood illustrates that the motion of the water is transverse (perpendicular) to the direction of motion of the wave. The distance between adjacent crests (or troughs) is the wavelength, λ.

move only in a localized region of space. In this case the air molecules vibrate back and forth along the direction of the wave; for this reason the wave motion is called a longitudinal wave. The similarity between the two types of wave motion just described is that in each case a disturbance travels through some kind of medium, be it air, water, or something else. Light is a transverse wave just like the water wave described above. The unique feature of light waves that distinguished them from all other types of wave motion is that no material medium is necessary for the propagation of light.

There exists an important, fundamental relation between the wavelength of a wave and its velocity. Consider a wave source that is vibrating periodically with a time interval T for one complete cycle. In the case of water waves the wave generator might consist of a ruler that is dipped repetitively into the water, causing equally spaced crests to move away from the wave source with a velocity V. The speed of propagation of the waves is given as

$$V = \lambda/T \tag{6-1}$$

Since the frequency f of periodic motion is related to the period T by $f = 1/T$, this relationship between velocity and wavelength can also be written as

$$V = f\lambda \tag{6-2}$$

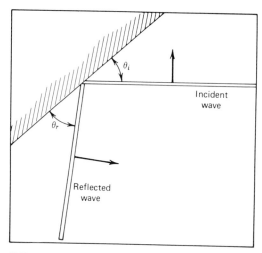

FIGURE 6-2 This diagram shows a straight wave moving toward a diagonal barrier from which it is reflected. The angle of incidence θ_i equals the angle of reflection θ_r. The direction of motion of the two rays is indicated by arrows.

This general equation is true for any simple, periodic wave; given any two of the three quantities V, f, and λ, the third is determined by this relationship.

Another useful property of waves can be obtained by observing what happens when they are reflected from some object. Consider the straight wave incident on a diagonal reflecting surface as shown in Figure 6-2. The angle that the incident wave makes with the reflecting barrier is labeled as θ_i; while that between the reflected wave and the barrier is shown as θ_r. Measurement of the angles θ_i and θ_r always gives that $\theta_i = \theta_r$. This relationship expressing the equality of the angles of incidence and reflection is known as the law of reflection. Using the law of reflection the focusing properties of any type of reflecting optics can be determined.

B. LIGHT

The type of wave motion of most interest to us is that of light. The sun, the stars, and electric lamps give off light. Other objects, such as the top of a desk, a refrigerator door, and this page do not. Whether a body emits light depends on its condition. In the next section we will find that this important condition of a material object is its temperature.

Light differs from other forms of wave motion in that no physical medium is necessary for the light to propagate. It was the vibration of the particles of the medium that generated the motion of sound and water waves. In the case of light the theory of electricity and magnetism shows that light is just one form of a general type of wave motion called electromagnetic waves. Electromagnetic waves are produced whenever electric charges are accelerated. In this case it is the electric and magnetic fields that vibrate. In free space (a vacuum) the velocity of all such waves is the same regardless of their frequency or wavelength. Following historical precedent it is customary to call electromagnetic waves of different frequency (or wavelength) by different names according to their origins. Figure 6-3 illustrates the different regions of the electromagnetic spectrum.

The radiation coming from our sun is of paramount importance to us. If we look back at Figure 5-5, this radiation can be usefully divided into three wavelength regions.

1. The short wavelength radiation at the left of Figure 5-5 is called ultraviolet (UV) radiation and is not visible to the human eye. These are the rays that are primarily responsible for a suntan. Most of the UV radiation that enters the solar atmosphere is absorbed there and does not reach the earth's surface.

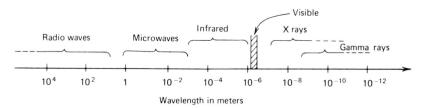

FIGURE 6-3 The electromagnetic spectrum is a continuous range of waves ranging from radiowaves to gamma rays. The descriptive names for sections of the spectrum are historical; they merely give a convenient classification according to the source of the radiation. The physical nature of the waves is the same throughout the entire frequency (wavelength) domain.

2. The middle wavelengths in Figure 5-5 are referred to as the visible spectrum, since these are the wavelengths that can be seen by the human eye. When this white light is passed through a specially constructed triangular piece of glass called a prism, the varying speeds at which these wavelengths travel cause it to be broken into colors that vary from violet through the blues and greens to the reds. The visible spectrum is an extremely narrow band when compared to the ultraviolet and the infrared bands.

3. The long wavelength is referred to as infrared radiation (IR), or heat radiation, although the latter term is not a good one since both ultraviolet and infrared will heat an object. All objects can be considered to be emitting infrared radiation, even a cake of ice. However, as we will soon see, the amount of energy radiated depends strongly on the temperature. Materials with a temperature below 800°F emit only in the infrared, with no visible or ultraviolet radiation. Thus, the radiation emitted from surface temperatures below 800°F is quite different in quality from the radiation of the sun.

A solar collector absorbs energy in all three of the wavelength regions. However, as can be seen from Figure 5-5, the energy at very long wavelengths will not be transmitted by the usual glass covers or envelopes used in most devices and is consequently not captured as useful heat. Fortunately, this wavelength region comprises only a small fraction of the energy emitted by the sun.

All bodies reflect some of the light that falls on them. In the case of a thin piece of glass only a small amount is reflected, since most of the light passes through the glass. This useful property of glass is utilized in the design of solar collector systems. An object, such as a thin sheet of glass or plastic, which passes most of the light incident on it, is said to be transparent. By contrast, light does not pass through a thin sheet

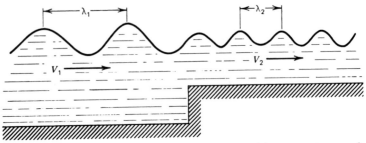

FIGURE 6-4 When water waves pass from a region of deep water to one of shallow water, their velocity of propagation is changed. An analogous phenomenon occurs when light passes from one medium to another. For a wave of a given frequency Eq. 6-2 shows us that the wavelength also changes.

of metal or wood. Materials with this property are termed opaque. Highly polished sheets of silver or aluminum will reflect most of the light incident on them. Reflecting mirrors are used in many different optical and solar energy applications to change the direction and intensity of the light.

When light passes into or out of a transparent material it often changes direction. This bending of the beam of light is termed refraction. The physical basis for this bending is due to the fact that the velocity of light depends on the medium through which it travels. For a wave of constant frequency this means that the wavelength will be changed as the velocity changes according to Eq. 6-2. This change in wavelength is illustrated in Figure 6-4 by water waves passing from a

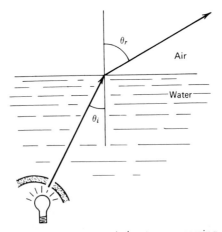

FIGURE 6-5 A wave is bent upon passing from one medium to the next where the velocity of propagation is different. In the case shown, a light wave travels more slowly in water than in air, so that a light ray in the water is bent away from the normal when it leaves.

medium defined by one velocity (in the water wave case, the velocity is determined by the wavelength and the depth of the water) to another where the wave velocity is different. Because of this change in velocity at the boundary, the direction of motion of the wave front will be different in the two media. This phenomenon is illustrated in Figure 6-5, which shows light being bent away from the normal upon passing from water to air. This phenomenon is termed refraction. Optical devices such as lenses make use of this property of wave motion.

C. BLACKBODY RADIATION

When a solid substance is heated it is found to emit radiation. A warm material emits radiation even though it often cannot be visually observed. As the temperature is increased, visible radiation that can be detected by eye is emitted. The hot material will go from "red hot" to "white hot" as the temperature is raised. The same phenomenon occurs for all objects whether solid or liquid. The energy radiated by a hot object is not monochromatic; instead it is distributed continuously over a wide range of wavelengths as indicated in Figure 6-6. As the temperature increases, more energy is associated with each wavelength region. In addition the peak of the distribution curve shifts toward shorter wavelengths. The similarity in shapes of the radiation curves for different temperatures led physicists to try and explain these phenomena with a single physical theory. Initially, this was made difficult by the fact that different distribution curves were obtained at the same temperature, depending on the nature of the emitting surface. However, it was soon found that at a given temperature, there was a maximum of radiation, at any specified wavelength, that could be emitted. By utilizing proper experimental techniques it was possible to obtain curves like those shown in Figure 6-6, corresponding to the maximum emission at a given temperature. When a heated object emits as efficiently as possible, regardless of the material of the emitting surface, it is called a *blackbody* emitter. The reason for this terminology will soon become apparent.

In order to understand how to utilize the radiant energy coming from the sun it is necessary to investigate the precise nature of the rules that govern solar radiation. The following discussion covers the three most important relations for blackbody radiation.

The most important relation is known as the Planck distribution function. Since its mathematical form is complicated, the exact functional form is not presented here. Instead, we just note that the curves in Figure 6-6 can be precisely fitted with the blackbody theory of Planck. Each curve on the figure represents a blackbody emitter of a different temperature. The curve for the sun, corresponding to 5800 K, is not shown as it would be far off the top of the page.

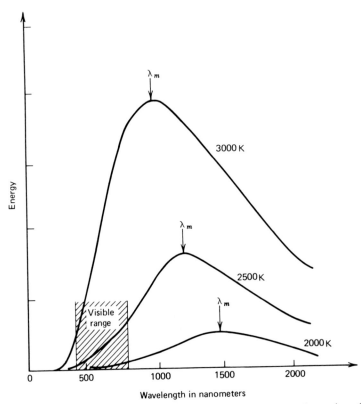

FIGURE 6-6 Distribution of energy emitted as a function of wavelength from a hot object. The solid lines show plots of the blackbody radiation curves for temperatures of 2000 K, 2500 K, and 3000 K. The wavelength at which the maximum amount of energy (per unit wavelength interval) is radiated is indicated by an arrow.

A careful study of Figure 6-6 can reveal some further interesting information. Any hot emitter emits radiation with a broad range of wavelengths. The peak of each curve corresponds to the wavelength at which the intensity is greatest for that particular temperature. We will come back to this point later. Note also that the higher the temperature of the blackbody emitter, the more its radiation is concentrated at the shorter wavelengths. It is apparent from these curves that the higher the temperature of the blackbody emitter, the larger the total amount of radiation emitted. This latter quantity is proportional to the area under each curve.

The above statements can be put into a useful quantitative form. That the amount of energy radiated depends strongly on temperature was first noted from the experimental data. In the late 1800s, after a long series of careful measurements of the total radiation emitted by a hot, black surface, Stefan showed that the energy radiated away from the surface was proportional to T^4, where T is the absolute temperature in degrees Kelvin. The same relation was derived theoretically by Boltzman using very general thermodynamic reasoning. This correspondence between the emitted radiative energy and the temperature of an object is known as the Stefan-Boltzman radiation law and is quantitatively expressed as

$$E = \sigma T^4 \tag{6-4}$$

where $\sigma = 5.67 \times 10^{-12}$ W/cm^2-K^4 and E is the total power emitted at all wavelengths.

Another important relation for solar energy collection concerns the position of the maximum in the curves of Figure 6-6. The wavelength corresponding to the maximum emission of radiant energy becomes smaller as the temperature increases. Using very general thermodynamic arguments, Wien showed that this maximum wavelength λ_m was quantitatively related to the temperature of the emitting object by

$$\lambda_m = \frac{\text{constant}}{T} \tag{6-5}$$

This relation, commonly called the Wien displacement law, is of fundamental importance in describing the trapping of solar energy by the greenhouse effect. The following example illustrates the effect of this basic law of radiation.

EXAMPLE 1

Assume that the sun is a blackbody emitter corresponding to a temperature of 5800 K. This gives rise to the solar radiation spectrum illustrated in Chapter 5 with the peak in the intensity of the radiation occurring at about 500 nm. Let us compare this peak wavelength for the radiation emitted by the sun with the peak wavelength emitted by the absorber plate of a solar collector, which typically operates at an average temperature of 140°F (330 K). Using Eq. 6-5,

$$\lambda_{5800} T_{5800} = \text{constant} = \lambda_{330} T_{330}$$

or

$$\lambda_{330} = 500 \text{ nm} \times \frac{5800 \text{ K}}{330 \text{ K}} \cong 8790 \text{ nm}$$

The blackbody distribution curve for the solar collector peaks at a wavelength far removed from the visible region, with the resulting emission consisting of radiation in the infrared region.

What has this to do with the greenhouse effect? Most solar collectors are covered with glass or plastic. Each of these covering materials is transparent to most of the light from the sun except for that small portion having very long wavelengths. Thus the incident solar intensity passes through the transparent covering of the collector and can be absorbed. On the other hand, the loss of energy from a collector is partly due to reradiation at very long wavelengths, as is seen in the example. This long wavelength radiation cannot pass through the collector covering, and is thereby trapped. This is not the whole story, of course, and the rest of this discussion is deferred to the next chapter.

Our next step is to determine exactly what is meant by a black surface. To do this, consider the reflection and absorption of radiation incident normally (perpendicular) on a thin surface as shown in Figure 6-7. Let the total radiation intensity incident on the surface be called I_0. The amount reflected is then rI_0, the amount transmitted is called τI_0, and the amount absorbed is called αI_0. Since the total intensity incident on the surface is either reflected, transmitted, or absorbed,

$$r + \alpha + \tau = 1 \qquad (6\text{-}6)$$

where r, α, and τ are the coefficients of reflection, absorption, and transmission.

Next, take the special case of a thick surface with $\tau = 0$, so that $r + \alpha = 1$. In this case the incident radiation is either absorbed or reflected. For a given material and a specific wavelength of incident radiation, r and α have definite values. If r is known, then $\alpha = 1 - r$ is also known. (Extending the discussion to the concept of radiation striking the surface obliquely would not change any of the conclusions.) It is now possible to define exactly what is meant by a white and black surface. A white surface has $r = 1(\alpha = 0)$ while a black one has $r = 0(\alpha = 1)$. That is, a black surface absorbs all of the radiation incident on it and reflects none, while the opposite is true for a white or perfectly reflecting

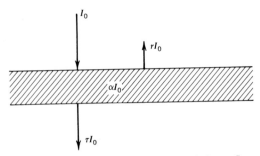

FIGURE 6-7 Schematic illustration of the reflection, absorption, and transmission of light through a thin slab of partially transparent material. The meaning of the symbols is given in the text.

surface. Surfaces corresponding to real objects are never perfectly white or perfectly black. Real surfaces also often have the property that the absorptivity α is not a constant, but instead varies as the wavelength changes. This variation of α with wavelength can be utilized to design more efficient solar collectors, as is shown in the next chapter.

It is now possible to show why an ideal emitter is called a blackbody emitter. Experiments show that as α approaches unity, the emissive power of a surface approaches that given by Eq. 6-4. Let us call the relative emissive power to that of a blackbody emitter its emissivity, ε. In 1859, G. R. Kirchoff showed that at any given temperature the emitting power of a surface is directly proportional to its absorbing power. Thus, a surface that absorbs all radiation of all wavelengths would also be the best possible emitter at all wavelengths; the total radiation from it would depend solely on its temperature and not at all on its chemical or physical nature. In equation form, Kirchoff's law states that at any given wavelength,

$$\varepsilon = \alpha \qquad (6\text{-}7)$$

A black object, previously defined as one that totally absorbed all the radiation incident on it ($\alpha = 1$), can also be defined as an object that emits radiation at the maximum rate possible ($\varepsilon = 1$). Most hot objects are not perfect emitters. The total radiation emitted by a real object is

$$E = \varepsilon \sigma T^4 \qquad (6\text{-}8)$$

FIGURE 6-8   The slot fire arrangement of L. Cranberg. The smallest logs are placed in front with larger logs on top and at the back. The net effect is to create a miniature blackbody emitter in the form of the cavity at front. The heat is partially trapped in the cavity resulting in a higher temperature there, and the amount of radiation to the room is increased.

The radiation emitted by a hot object is reduced by the factor ε, which is always less than 1.

The concepts of blackbody radiation have been utilized recently to design a better fireplace fire, as illustrated in Figure 6-8. The standard fireplace grate is supplemented by two metal uprights at the front corners that are fitted with adjustable arms extending into the fireplace. A large log is then placed at the rear of the grate with smaller ones at the front. Another large log is placed on the moveable arms, which are adjusted so that the upper log just contacts the one in the rear of the fireplace. This creates a cavity that opens into the room; about 30% of the heat generated there is radiated into the room. A very important bonus is that the fire is easy to light. Because this arrangement traps heat so well, damp wood can be lit with only a few sheets of newspaper placed directly in the cavity.

D. MEASUREMENT OF SOLAR RADIATION

The purpose of solar instruments is to measure the energy associated with radiation incident on a plane of given orientation and to provide information about the spectral and spatial distribution of this energy. These instruments convert the energy of the incident solar radiation into another form of energy that can be measured more conveniently.

The most commonly used instruments for determining the available solar energy are pyrheliometers for measuring the radiation coming directly from the solar disk, pyranometers for measuring the combination of direct solar radiation and that scattered by particles and molecules in the atmosphere, and duration of sunshine instruments for measuring the fraction of daylight hours during which the view of the sun is not blocked by clouds.

It is not the purpose of this text to present a detailed discussion of solar instruments. The problems of calibration, the methods of converting solar radiation into electrical signals, the techniques used to compensate for changes in ambient temperature, etc. are described elsewhere in technical books on solar radiation. It is important here only that we have an idea of the basic elements of the two most common monitoring instruments, the pyranometer and the pyrheliometer.

The pyranometer is used to determine the direct and diffuse radiation together, that is, it measures global radiation. One type of design is shown in Figure 6-9. The receiver consists of two concentric rings. The outer ring is white and the inner ring is black. The rings are often made of silver foil, 0.25 mm thick, and the coatings are magnesium oxide and Parson's black. A thermal insulator is inserted between the rings. The dull, black surface absorbs almost all the radiation incident on it, while

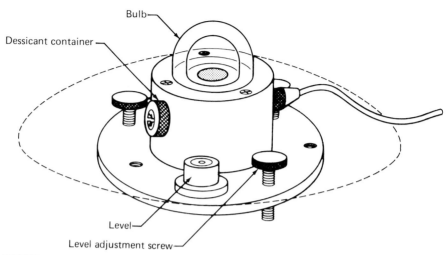

Bulb

Dessicant container

Level

Level adjustment screw

FIGURE 6-9 Pictorial representation of a pyranometer. The light absorbed by the black absorber develops a temperature difference between it and the surrounding world. This temperature difference is proportional to the energy absorbed.

the white magnesium oxide reflects visible and near-infrared radiation. The energy absorbed by the receiver generates a temperature difference between the rings, and this temperature difference is then measured by a number of thermocouples.

The pyranometer is used to measure the total radiation incident on a given surface (usually horizontal). With good instruments such as the Eppley pyranometer, experiments have shown the absence of spectral selectivity, since the glass transmits practically all radiation between 350 and 20,000 nm. The response is quick enough that 98% of the maximum output is reached in 20 to 30 seconds. External influences are kept to a negligible level by sealing the instruments. The principal factors affecting the calibrations are temperature, angle of incidence of the sun's rays (often called the cosine correction), and the tilt of the pyranometer. The temperature correction can be as large as ±5% for ±30°C changes for poorer quality instruments. Similar uncertainty often occurs for the cosine correction as the zenith angle of the sun (angle measured from the vertical) approaches 90°.

The direct beam from the sun is measured with an instrument called a pyrheliometer. This instrument is constructed with a receiving element like that for a pyranometer, and the directional sensitivity is achieved through the use of a long collimator (Figure 6-10). To reduce errors arising from imperfect alignment with the sun, pyrheliometers are constructed to include a part of the radiation from the sky immediately around the sun. The angular acceptance of a pyrheliome-

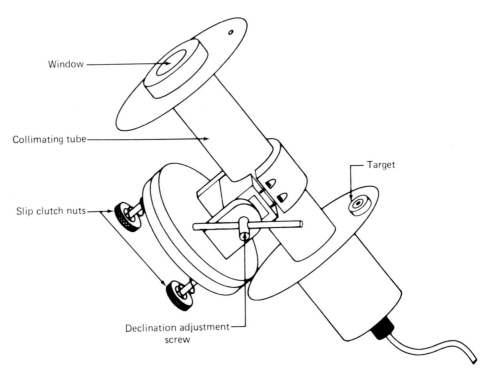

Window

Collimating tube

Slip clutch nuts

Target

Declination adjustment
screw

FIGURE 6-10   Illustration of the geometry used for a pyrheliometer. This instrument accepts radiation in a cone making a full angle of 6°. The sensing element at the bottom can be similar to that used in a pyranometer. The walls of the collimating tube are painted black to eliminate scattered light.

ter is determined by the dimensions of the cylindrical tube and associated diaphragms. Typically, the instrument is designed so that the angle subtended by any point on the receiver is 5°, 41 min. The pyrheliometer is mounted on a polar tracking mechanism similar in principle to that used for astronomical telescopes. The pyrheliometer then rotates 360° every 24 hours and minor adjustments of the polar and azimuthal angles of the solar instrument during the year are sufficient to keep it in alignment.

E.   THE INCIDENT SOLAR RADIATION

Although the earth intercepts only a small fraction of the total radiation emitted by the sun, the amount received per year is equivalent to tens of thousands times the present annual energy requirement for the world. Over the period of a year a square mile receives about 1.3×10^{16} J. The amount received per year varies considerably with geographical

TABLE 6-1
Monthly Average Daily Solar Radiation[a]

SITE	JAN.	FEB.	MAR.	APR.	MAY	JUNE	JULY	AUG.	SEPT.	OCT.	NOV.	DEC.	TOTAL
Albuquerque, NM	11.54	15.23	20.06	25.29	28.80	30.40	28.24	25.99	22.38	17.55	12.87	10.53	7579
Atlanta, GA	8.14	11.00	14.79	19.14	21.04	21.72	20.57	19.39	16.14	13.62	10.02	7.65	5574
Boston, MA	5.40	8.05	11.53	15.05	18.39	20.62	19.85	16.87	14.30	10.10	5.71	4.57	4577
Chicago, IL	5.75	8.62	12.56	16.56	20.30	22.78	22.06	19.51	15.36	11.00	6.42	4.56	5033
Denver, CO	9.53	12.79	17.37	21.33	24.73	26.68	25.79	23.20	19.60	14.76	10.03	8.30	6497
Hartford, CT	5.42	8.11	11.10	14.92	17.80	19.13	18.71	16.13	13.10	9.68	5.64	4.37	4384
Kansas City, MO	7.35	10.15	13.65	17.87	21.25	23.60	23.86	21.14	16.48	12.40	8.37	6.37	5552
Miami, FL	12.00	14.91	18.20	21.10	20.92	19.38	20.01	18.50	16.53	14.78	12.67	11.57	6103
Minneapolis, MN	5.27	8.67	12.52	16.36	19.72	21.88	22.36	19.15	14.24	9.76	5.45	4.01	4847
New Orleans, LA	9.47	12.62	16.06	20.20	22.33	22.74	20.58	19.48	17.18	15.15	11.04	8.84	5953
Phoenix, AZ	11.59	15.59	20.59	26.72	30.37	31.09	28.22	26.02	22.87	17.89	13.06	10.58	7745
Portland, ME	5.11	7.74	11.00	14.80	17.79	19.42	18.83	16.58	13.14	9.33	5.21	4.12	4352
San Francisco, CA	8.03	11.45	16.51	21.79	25.26	26.97	27.14	24.02	19.77	13.91	9.32	7.29	6481
Seattle-Tacoma, WA	2.97	5.62	9.64	14.68	19.45	20.45	25.51	18.34	13.02	7.45	3.83	2.40	4368
Tucson, AZ	12.47	16.25	21.16	26.82	30.32	30.98	26.57	24.77	22.46	18.18	13.71	11.30	7756

[a] All monthly average values are in MJ/m^2 per day. The last column gives the annual radiation in MJ/m^2.
Source. The data base for this table is from the report "Input Data for Solar Systems" by V. Cinquemani, J. R. Owenby, Jr., and R. G. Baldwin of the National Climatic Center, Asheville, North Carolina.

region, as shown in Table 6-1. Knowledge of the incident solar flux is clearly needed for most solar energy applications. The intensity of the available solar radiation—the amount of energy per unit area and unit time received on the earth—varies considerably with the season, the amount of water vapor and other particles in the atmosphere, the solar altitude, and the orientation of the receiving surface. The one fixed quantity is the solar radiation intensity outside the earth's atmosphere.

The total solar energy integrated over all wavelengths received per unit time by a unit area of surface oriented normally to the sun's rays at the top of the earth's atmosphere is called the *solar constant*. The radiation incident on this surface is illustrated in Figure 6-11. Since by definition this surface is always oriented perpendicular to the incoming solar radiation, the amount of energy received by a given area of surface should be independent of the time of year. This is the reason for calling the energy received per unit area per unit time the solar constant. The determination of this constant has been the object of a number of experimental measurements over the years.

The earliest determinations, made from high mountains, had to be corrected for the transmission through the atmosphere of the different wavelength portions of the solar spectrum. More recent measurements with high-altitude aircraft, balloons, and spacecraft have permitted direct determinations outside most of the earth's atmosphere. The best recent value for the solar constant is 1370 W/m^2 (1.962 cal/cm²-min, 434.5 Btu/ft²-hr). The absolute uncertainty in this constant has improved considerably over the past 10 years and is now probably only

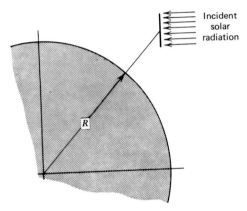

Incident solar radiation

FIGURE 6-11 The solar constant is the amount of solar energy incident per unit time on a surface of unit area oriented perpendicular to the sun's rays at the top of the earth's atmosphere.

± 5 W/m^2. These recent precise absolute measurements also indicate short-term intensity variations of the order of ± 3 W/m^2, perhaps because of changes in sunspot activity.

It is more precise to say that the solar constant as defined above is determined at the average distance between the sun and the earth. The orbit of the earth around the sun, though close to circular, is actually an ellipse. This causes a variation of the solar intensity of the order of $\pm 3.3\%$. The maximum intensity on a surface oriented perpendicular to the sun's rays occurs in December, reaching 1.033 times the average

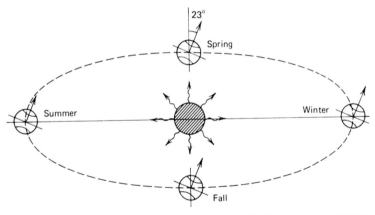

FIGURE 6-12 Graphical illustration of the reason for the changing seasons. Because of the tilt of the earth's axis, the amount of solar radiation incident on a horizontal surface changes throughout the year.

value (the solar constant). The minimum value occurs in June when the intensity is 0.967 times the average value.

It is important to understand the reason for the changing seasons as the earth makes it's passage around the sun once a year. As shown in Figure 6-12, the seasonal variation is caused by the fact that the earth rotates around an axis that is inclined at 23.5° to the perpendicular to the plane of the earth's orbit. What is important in this case is the solar intensity on an area parallel to the ground. In the northern hemisphere the maximum period of daylight is at June 22, while the minimum period of daylight occurs at December 22. The day and night periods are of equal length at the equinoxes, which come at March 21 and September 23.

F. SOLAR INTENSITY AT THE EARTH'S SURFACE

Almost half the solar radiation striking the earth's upper atmosphere is either absorbed by air molecules or reflected and scattered by clouds and small particles before reaching the surface. The intensity of the solar radiation striking the earth's surface thus depends on the amount of air mass between the observer and the sun and on the amount of water vapor, dust, ozone, and other molecules present in the atmosphere. The air mass is clearly greater the lower the sun's position in the sky, and is a minimum for a given day at solar noon.

In principle, we could start with the solar constant and obtain the solar radiation intensity at ground level by correcting for the attenuation in intensity due to the air mass and cloud cover of the atmosphere. This approach is seldom useful in practice because of large uncertainties in making the corrections, particularly that for the attenuation due to clouds. The most reliable way to determine the solar intensity on the ground is by direct measurement. Measurements of the intensity of solar radiation are usually recorded as total radiation on a horizontal surface. This quantity, also often called global radiation, is measured with a pyranometer.

The total solar radiation at the ground consists of the following three components:

1. Direct (or beam) radiation, which comes straight from the sun, as shown in Figure 6-13.

2. Diffuse (scattered) radiation, which comes from the entire sky. As illustrated in Figure 6-13, this component arises from scattering of the incident solar radiation by particles or molecules in the

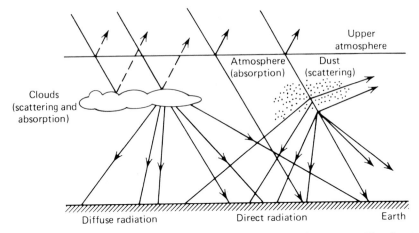

FIGURE 6-13 The incoming solar radiation is composed of two parts. The direct beam consists of radiation coming straight from the sun. The diffuse component is made up of radiation scattered one or more times by the various constituents of the earth's atmosphere.

atmosphere. On a very clear day, the diffuse component may amount to only 10 to 20% of the total intensity, while on overcast days the diffuse fraction will approach 100%.

3. Reflected radiation, which comes from nearby surfaces such as buildings or the ground.

The important factors affecting the observed intensity at ground level are scattering by air molecules, water vapor, and dust and the atmospheric absorption by such gases as N_2, O_2, O_3 (ozone), H_2O, and CO_2. Looking first at air molecules, we note that they are very small compared to the wavelengths of radiation significant in the solar spectrum. This type of scattering of light was first investigated by Lord Rayleigh, who showed that the scattered intensity was given as

$$I_\lambda(\theta) = I_{0\lambda}(1 + \cos \theta^2)/\lambda^4 \qquad (6-9)$$

where θ is the angle between the direction of incidence and the direction of scattered light and $I_{0\lambda}$ is the intensity of light incident upon the molecule. Maximum scattering occurs in the forward and backward directions. Even more important is the dependence on the wavelength λ. Since the wavelength of blue light is considerably smaller than that of red light, the blue light is scattered much more than the red wavelengths. For example, a factor of 2 difference in wavelength would result in a factor of 16 more scattering for the short-wavelength radiation. This provides a natural explanation of the blue color of a clear sky as illustrated schematically in Figure 6-14.

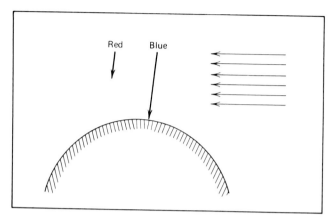

FIGURE 6-14 The enhanced scattering of light of short wavelength (blue) results in the bluish color of a clear sky.

Scattering by dust particles and water vapor is more difficult to assess, since their size is much larger than that of air molecules. Empirical equations have been developed to give an approximate correction for this type of scattering. While the scattering from a single, large particle is often in the forward direction, a thick cloud layer will result in multiple scattering of the incident light so that the observed intensity appears to come equally from the entire sky.

In addition to scattering, the atmosphere also absorbs some of the incident solar radiation. Absorption of short-wavelength radiation (ultraviolet rays) is due primarily to ozone (O_3). Most of this ultraviolet absorption occurs in thin layers of ozone located quite high in the atmosphere; the destruction of this ozone layer would allow the ultraviolet radiation to reach the earth's surface. There is only a small amount of absorption in the visible region. Absorption of the longer wavelength radiation (infrared) is due principally to molecules of water and carbon dioxide. The result of the various interactions with the atmospheric constituents is to reduce the intensity of solar radiation at the earth's surface to a value considerably below that of the top of the atmosphere.

G. DIRECT AND DIFFUSE COMPONENTS

An attractive feature of a simple, flat-plate collector for water and space heating is that it will collect both the direct and diffuse components of solar radiation with almost equal efficiency. This is important for these common, low-technology applications because the only region where the influence of clouds on solar radiation is negligible is in the desert. If only the diffuse component were important then the optimum orientation of a flat-plate collector would be horizontal so that it could be exposed to the entire sky. Nevertheless, a considerable fraction of the incident radiation is direct, and for this component a significant improvement in the radiant energy received can be achieved by tilting the collector toward the vertical without much loss in the collection of the diffuse component. Because of this the first step in estimating the amount of collectible solar energy is to somehow separate out the direct and diffuse components of the incident solar intensity. As noted earlier, in principle we could obtain the direct and diffuse components of the incident solar radiation by an involved calculation starting from our knowledge of the solar constant. In practice it is difficult to know the precise composition of the atmosphere. Clouds are a complex phenomenon, ranging from thin, transparent cirri that attenuate the incident light only slightly to thick, thunderstorm clouds that can reduce the solar radiation to as low as 1 to 2% of its upper-atmospheric value. There are very few situations for which the type and extent of each cloud layer is observed, allowing a direct calculation of the ground level radiation.

The ideal method is to measure the direct component with a pyrheliometer and the total intensity with a pyranometer. The diffuse component can then be obtained by making a simple geometrical projection of the direct component on a horizontal surface and subtracting from the total intensity. This procedure is followed at the very best solar radiation monitoring sites, but requires a considerable investment in equipment and personnel.

In practice the following simplified procedure is often followed. Considerable information is available for the total solar intensity on horizontal surfaces. The fraction of diffuse radiation can then be estimated in a statistically significant way by utilizing a procedure first developed by Liu and Jordan of the University of Minnesota during the early 1960s. They developed an empirical procedure for estimating the percentage of diffuse and direct components, given only the average total radiation. To do this they introduced a parameter called the clearness index, K_T. Formally, the clearness index is defined as the ratio of the measured, total radiation upon a horizontal plane of the earth's surface to the corresponding quantity at the top of the atmosphere. That is,

$$K_T = \Phi/\Phi_0 \qquad (6\text{-}10)$$

where Φ_0 is the solar intensity on a horizontal surface at the top of the atmosphere, which can easily be calculated from a knowledge of the solar constant. This parameter was then empirically correlated with the diffuse component obtained from data from recording stations that also recorded both the direct and total solar radiation. From a statistical analysis of this correlation, Liu and Jordan discovered that a firm

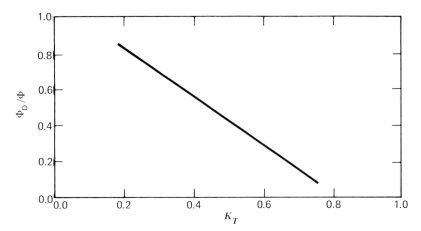

FIGURE 6-15 Plot of the empirical relation between the ratio of the diffuse component to the total solar intensity and the clearness index, K_T. All solar radiation values are monthly averages.

relationship existed between K_T and the diffuse component. This relationship is graphically illustrated in Figure 6-15 for the case where monthly averages are used in the analysis. The highest plotted values of K_T are near 75 to 80%, indicating that under total clear-sky conditions, about 20 to 25% of Φ_0 is lost as a result of scattering and absorption by the atmospheric air mass.

At the other end of the graph where K_T is about 5%, the diffuse percentage is about 100%. In this case almost no direct sun is seen; the cloud cover required for this condition is very dense—perhaps 4 to 5 miles deep. The relation outlined in Figure 6-15 is not strongly dependent on latitude or other factors peculiar to a particular location. Using this graph, the monthly average daily diffuse and direct solar components can then be estimated from knowledge of the monthly average total solar radiation.

H.   RADIATION ON A TILTED SURFACE

It is the solar intensity on a horizontal surface that is commonly measured, but most collector systems are tilted away from the horizontal to increase the amount of solar energy collected. The reason for this is simple. The amount of direct radiation captured is maximized when the receiving surface is oriented so that it is perpendicular to the incoming direct radiation. As is shown below, the amount of direct radiation on a tilted surface can be obtained using simple geometry.

Before doing this let us see how the position of the sun in the sky is described. This is illustrated in Figure 6-16. The two angles θ_T and ϕ

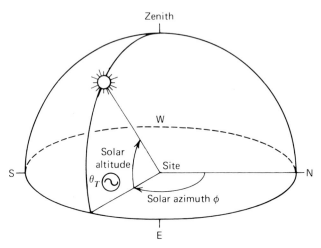

FIGURE 6-16 The position of the sun in the sky is defined in terms of the solar altitude angle θ_T, and the azimuth angle ϕ.

suffice to describe the sun's position in the sky. The solar altitude angle θ_T is the angle measured vertically from the horizon to the sun. It equals 0° when the sun is on the horizon (sunrise or sunset) and 90° when the sun is directly overhead. Solar azimuth is the angular distance of the sun from true north. For example, $\phi = 180°$ at true south. (For many technical purposes, the azimuth angle is set to zero degrees at the point when the sun points true south to an observer in the northern hemisphere.) In what follows it will prove more convenient to use the polar angle θ, which is defined as $90° - \theta_T$ and is shown in the upper portion of Figure 6-17.

It is now easy to calculate the amount of direct solar energy incident on a tilted surface in terms of the corresponding direct quantity measured (or obtained as described in the previous section) on a horizontal surface. If the intensity of the incident direct beam is I_0, then the corresponding intensity of the solar radiation on a horizontal surface is just $I_H = I_0 \cos \theta$. If the collector surface is tilted then the intensity is $I_{\text{DIR}}(\theta_T) = I_0 \cos \theta_S$, where $\theta_s = \theta - \theta_T$. If $\theta_T = \theta$, then the intensity on the tilted surface achieves the maximum value I_0. This minor bit of geometry illustrates the obvious fact that the maximum amount of solar radiation is received when the collector surface is oriented to be perpendicular to the incident solar radiation. The amount of solar radiation received by a tilted surface can then easily be related to that obtained for a horizontal surface:

$$I_{\text{DIR}}(\theta_T) = I_H \frac{\cos \theta_S}{\cos \theta} = I_H \frac{\cos (\theta - \theta_T)}{\cos \theta} \qquad (6\text{-}11)$$

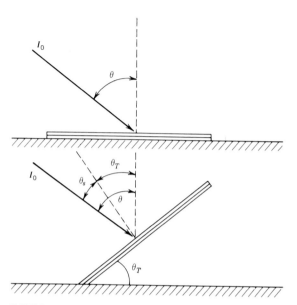

FIGURE 6-17 Showing the geometrical situation for direct solar radiation. The top portion of the figure shows direct rays incident on a horizontal surface at an incident angle θ. When a tilted surface at angle θ_T is used, the incident radiation now makes an angle $\theta_s = \theta - \theta_T$ with this surface.

The angle θ can be calculated from a knowledge of the latitude and the time of day. The thoughtful student will recognize that for the computation of the solar radiation received over the entire day, the angle θ will change with time, and a weighted average must be used to convert the total daily radiation measured on a horizontal surface to the corresponding value on the tilted one.

The amount of the diffuse component of solar radiation received by the solar collector also varies with tilt angle. The calculation of this fraction, assuming insolation to come equally from all points in the sky (called an isotropic distribution of solar radiation), is reasonably complicated, requiring a summation over that portion of the sky seen by the collector and including a correction for nonnormal angles of incidence. Although the calculation is tedious, the final result is simple:

$$\frac{I_D(\theta_T)}{I_D(\text{HOR})} = \tfrac{1}{2}(1 + \cos \theta_T) \qquad (6\text{-}12)$$

where $I_D(\theta_T)$ is the amount of diffuse radiation incident on a surface tilted to the horizontal by the angle θ_T.

The total amount of solar radiation incident on a tilted surface can

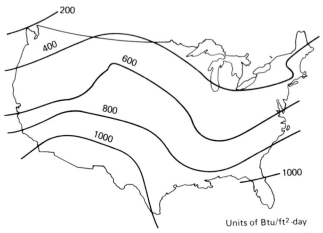

FIGURE 6-18 Average solar radiation for the United States in the month of December as measured on a horizontal surface. The solid lines indicate regions of equal incident radiation.

now be easily calculated using the above results. The direct intensity is calculated using Eq. 6-11, and the diffuse intensity is obtained from Eq. 6-12. A small correction is also added in for the intensity of reflected radiation.

When we use the above techniques it is interesting to compare the pattern of solar radiation in the United States on horizontal and tilted surfaces. The geographical distribution of radiation on a horizontal surface for the month of December is presented in Figure 6-18. Lines of

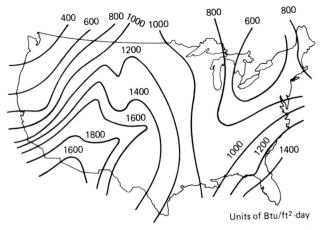

FIGURE 6-19 Average solar radiation for the United States in the month of December as computed on a tilted surface following the procedures outlined in the text. The solid lines indicate regions of equal incident radiation. Notice the large difference in the radiation pattern between this figure and the previous one.

equal solar intensity are shown by heavy, dark lines. It is seen that the radiation received generally falls with increasing latitude; the average daily total insolation in portions of the state of Washington is only about one-fifth that incident on New Mexico or Arizona (see also Table 6-1). By separating the total radiation data of Figure 6-18 into its direct and diffuse components and correcting to a tilted surface, the map shown in Figure 6-19 is obtained. The advantage of plotting the data in this manner is that they are now representative of the actual radiation on a real solar collector. An examination of Figure 6-19 shows that the pattern for direct radiation on a tilted surface is distinctly different from that obtained for the solar radiation on a horizontal surface. The Midwestern states are an area of high input with most of North Dakota and Montana (48° latitude) receiving more direct insolation than Washington, D.C. (39° latitude). The important conclusion is that for great portions of the central regions of the country, the solar radiation incident on a tilted surface during the winter is roughly independent of latitude.

REFERENCES

1. *College Physics, Physical Science Study Committee* (Bedford, Md.: Raytheon Education Company, 1968).

     Chapters 3, 4, 6, 7, and 9 of this elementary text provide a detailed, but very readable account of wave phenomena and light.

2. B. Y. H. Liu and R. C. Jordan, "The Interrelationship and Characteristic Distribution of Direct, Diffuse and Total Solar Radiation," *Solar Energy* 4,1 (1960).

     Presents relations permitting the determination on a horizontal surface of the instantaneous intensity of diffuse radiation on clear days, the long-term average hourly and daily sums of diffuse radiation, and the daily sums of diffuse radiation for various categories of days of differing degrees of cloudiness.

3. F. Y. Kondratyev, *Radiation in the Atmosphere* (New York: Academic Press, 1969).

     A thorough description of radiative processes in the earth's atmosphere, published as part of the international Geophysics Series. This book was written in English by Professor Kondratyev. Subjects covered include the measurement of solar radiation, absorption and scattering in the atmosphere, direct and diffuse components, energy distribution in the spectrum of global radiation, as well as many other topics. This book contains about 1500 references to relevant literature, although a considerable fraction of them are Russian papers.

4. *Sunworld*, Volume 4, pp. 75–115, 1980 (Oxford, England: Pergamon Press).

This special issue on solar radiation contains many readable articles on some of the topics covered in this chapter. Presents an elementary treatment of the subject along with several interesting historical reviews.

5. Kinsell L. Coulson, *Solar and Terrestial Radiation* (New York: Academic Press, 1975).
   A more advanced textbook concerned primarily with solar instrumentation.

QUESTIONS

1. When a rock is dropped in a large pool of water, a circular wave front is observed to move out radially. The water molecules:
   (a) Vibrate horizontally along the direction of the motion of the wave.
   (b) Vibrate in a vertical manner perpendicular to the wave front motion.
   (c) Move along with the wave front
   (d) Do not really move at all.

2. The wavelength of which of the following is the largest?
   (a) Light waves
   (b) Radio waves
   (c) X-ray waves
   (d) Gamma ray waves

3. When light waves pass from air into glass, the ray describing the motion of the wave front changes direction according to the law of refraction. The light ray is (Hint: Look at Figure 6-5):
   (a) Bent toward the normal to the glass surface.
   (b) Bent away from the normal to the glass surface.
   (c) Not bent at all, that is, the direction of the wave front is unchanged.
   (d) Light is completely reflected at the air-glass interface.

4. Suppose that we have a source of sound waves vibrating at a rate that gives a wavelength of 10 m in air. If the vibration rate is then reduced to one-half of the original rate, the new wavelength is:
   (a) Double the original value.
   (b) The same.
   (c) One-half of the original value.
   (d) Four times the original value.

5. A transverse wave is one for which the medium vibrates:
   (a) In a direction that makes a 45° angle with the direction of motion of the wave.
   (b) In a manner resulting in complete polarization of the wave.
   (c) In a direction parallel to the direction of the motion of the wave.

(d) In a direction perpendicular to the direction of the motion of the wave.

6. Suppose we have an opaque surface that reflects 95% of the far-infrared radiation incident on it. Its emissivity for this radiation is:

(a) 5%   (c) 90%
(b) 20%   (d) 95%

7. A pane of glass reflects 6% of the incident light and absorbs 12%. The amount of light transmitted is:

(a) 94%   (c) 82%
(b) 88%   (d) 18%

8. As the temperature of a hot, glowing object increases, the wavelength, λ_m, at the maximum of the blackbody radiation curve:

(a) Decreases.   (c) Remains the same.
(b) Increases.   (d) Makes a quantum jump.

9. A blackbody absorbs all of the incident radiation. It also:

(a) Reflects radiation with 50% efficiency.
(b) Emits radiation with 0% efficiency.
(c) Reflects radiation with 100% efficiency.
(d) Emits radiation with 100% efficiency.

10. The emissivity of an object is found to be 0.87 for radiation of a given wavelength. What is its reflectivity (assume $\tau = 0$)?

(a) 1/0.87   (c) 0.13
(b) 0.87   (d) 1/0.13

11. When visible light is scattered by air molecules, which wavelengths are scattered the least?

(a) Red   (c) Blue
(b) Yellow   (d) Violet

12. Which of the following constituents of the earth's atmosphere is mainly responsible for absorbing the ultraviolet radiation?

(a) Ozone   (c) Nitrogen
(b) Oxygen   (d) Water vapor

13. Compare the wavelength λ_1, at the point of maximum intensity of a blackbody spectrum from a system at a temperature of 6000 K, with the corresponding value λ_2, for a system at a temperature of 3000 K.

(a) $\lambda_2 = 2\lambda_1$ (c) $\lambda_1 = 2\lambda_2$
(b) $\lambda_2 = \lambda_1$ (d) Insufficient information

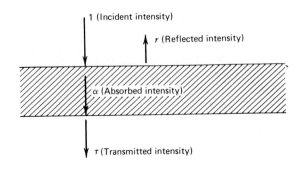

For the following three questions refer to the above figure. In general, $r + \alpha + \tau = 1$. (This is only meant to be a schematic illustration of the actual situation.)

14. Suppose that $\tau = 0$ (for a very thick piece of material). If the material is perfectly black:

(a) $r = 0$ (c) $\alpha + r = 0$
(b) $\alpha = 0$ (d) $2\alpha + r = 1$

15. Suppose that $\tau = 0$ as in the above problem. If the material is perfectly white:

(a) $r = 0$ (c) $\alpha = 0$
(b) $\alpha = 1$ (d) $\alpha + r = 0$

16. Suppose that we have an ideal piece of glass with $\alpha = 0$. If $r = 0.08$:

(a) $\tau = 0.08$ (c) $\tau = 1$
(b) $r + \tau = 1$ (d) $r + \tau = 0$

17. Which of the following wavelengths is scattered the most by air molecules. Waves with a wavelength of (one micron $= 10^{-6}$ m):

(a) 0.5 micron   (c) 0.7 micron
(b) 0.6 micron   (d) 0.8 micron

18. The intensity of the solar radiation on a surface oriented perpendicular to the sun's rays (at all times) which is located at the top of the earth's atmosphere is largest for which of the following days?

(a) June 21   (c) December 21
(b) September 21   (d) March 21

19. The intensity of the solar radiation on a fixed surface located at the top of the earth's atmosphere and oriented parallel to the horizontal plane is largest for which of the following days (Fixed surface located in the northern hemisphere)?

(a) June 21   (c) December 21
(b) September 21   (d) March 21

20. Suppose that you lived in a climatic region at 45°N latitude where clouds obscured the sky almost all of the time during the winter heating season. Assume that, under these conditions, the solar radiation was entirely diffuse. Which of the following would be the optimum orientation for your solar collector?

(a)   90°   (c)   30°

(b)   60°   (d)   0°

21. Suppose that the sun is located at 30° above the horizontal plane at noon on a certain day in January. Consider a flat piece of metal with an area of 4 m² that is oriented so that it faces south. For which of the following tilt angles (measured from the horizontal) will the amount of solar radiation received by the piece of metal be the largest?

(a)   0°   (c)   60°

(b)   30°   (d)   90°

22. A pyranometer measures which of the following?

(a)   Only the diffuse component of the solar radiation.

(b)   Only the direct component of the solar radiation.

(c)   The total amount of solar radiation incident upon the instrument.

(d)   The solar radiation received by the instrument in a cone with a half-angle of 6°.

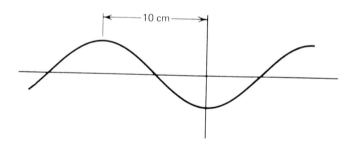

23. The above figure illustrates a waveform for a water wave moving to the right at 100 cm/sec.

(a)   Name the two fundamental types of wave motion.

(b)   Which of the two basic types of wave motion is the water wave?

(c)   What is the wavelength λ of the water wave?

(d)   What is the general relation between velocity, wavelength, and frequency for simple wave motion?

(e)   What is the frequency of the above wave?

(f)   If the frequency of the above wave is doubled, what will be the new wavelength in cm?

(g)   For a water wave it is the water which vibrates, giving rise to the wave motion. What is vibrating in the case of light or radio waves?

(h)   What vibrates in the case of the above water wave?

24.   Match up the items on the left with the most reasonable quantity on the right. (Write the appropriate letter)

_____ 1.   Water wave
_____ 2.   Sound wave
_____ 3.   Light
_____ 4.   Relation between frequency, wavelength and velocity of a wave
_____ 5.   Measurement of direct radiation
_____ 6.   Measurement of total solar radiation
_____ 7.   Infrared emissivity of a good selective surface
_____ 8.   Frequency of radio waves as compared to light waves

(a)   Less than
(b) $V = f/\lambda$
(c)   Pyranometer
(d)   Electromagnetic wave
(e)   Longitudinal wave
(f)   0.94
(g)   Greater than
(h)   Ratio of the velocity of light in glass to that in air
(i) $V = f\lambda$
(j)   Transverse wave
(k)   0.10
(l)   Pyrheliometer
(m)   Ratio of the velocity of light in air to that in glass

25.   What constituent of the atmosphere is responsible for absorbing part of the infrared portion of the spectrum?

26.   Why is a flat-plate solar collector better than the concentrating type of collector on cloudy days?

27.   A sound wave is sent down a long metal bar and eventually dies out. What has happened to its energy?

28.   The figure on the next page shows solar radiation incident normally on a pane of glass covering a solar collector.

(a)   Suppose that 4% of the incident radiation is reflected at each air-glass interface (boundary). What percentage of the incident radiation will be reflected by the pane of glass?

(b)   Suppose that the pane of glass in part (a) also absorbs 8% of the incident solar radiation. What percentage of the incident radiation will be transmitted?

(c) To improve the performance of the solar collector, the above pane of glass is replaced with one containing no iron. This pane of glass absorbs only 2% of the incident solar radiation. What percentage of the incident solar radiation is now transmitted?

(d) Suppose that we treat the surfaces of our iron-free glass with an antireflection coating. In this case the reflection losses at an interface are reduced to 2%. What percentage of the incident energy is now transmitted?

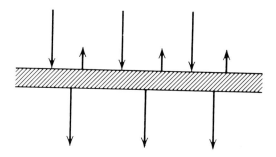

29. In the following problem you are to compare the item on the left with that on the right. Put a mark in front of the one that is largest. An example is included to clarify the procedure.

_____ Your house	_X_ Empire State Building
_____ Wavelength of light	_____ Wavelength of radio waves
_____ Velocity of light in air	_____ Velocity of light in glass
_____ Amount of light reflected by a mirror	_____ Amount of light reflected by a pane of glass
_____ Amount of light reflected by a pane of glass for normal incidence	_____ Amount of light reflected by a pane of glass with angle of incidence equal to 60°
_____ Frequency of a radio wave	_____ Frequency of light wave

30. When you stand in front of a mirror, is the image of yourself exactly like the view that others get of you? Describe the difference, if one exists.

31. You observe a rainbow. Describe the relative positions of the sun, the rain drops in the rainbow, and yourself.

32. Suppose that you are in a boat and are rowing toward a light source that you know to be submerged 2 m below the surface of the water. Suddenly you observe the light. Immediately you stop and turn around, rowing away from the light source. Soon the light source disappears from view. Knowing only that light rays bend away from the normal when passing from water to air, explain the observed phenomenon.

7 passive solar applications

An attractive approach to solar energy is to use the entire building as the collector and storage system. This type of structure is arranged to work in harmony with the local climate. When the sun shines the building gains heat and stores it for later use. With a passive solar system heat moves entirely by natural means. The transport of heat from one place to another is accomplished by the natural physical processes of conduction, convection, and radiation. Strictly speaking this would rule out the use of such devices as small fans or motorized insulation panels. Nevertheless, these items are often used in "hybrid" passive designs. But the goal of good passive solar design is to maximize the use of natural methods for the transport of heat.

This type of approach is hardly a new idea. Some of the early Southwest native civilizations practiced such designs; Western ranch houses have often adoped important features such as a southward exposure with long overhangs for summer shade, low profiles, and heat-efficient local building materials. The renewed interest in solar energy has also spurred the development of passive systems.

With passive systems, the method of collection of solar energy is easy, since one just orients the house due south with a large window exposure. The calculation of solar gains through window areas is well understood. The exact amount of usable heat gained depends on climatic conditions, the latitude, and the amount of insulation used. Properly oriented windows, combined with a means of insulation controlling heat losses during sunless hours, can often provide a large fraction of the annual space-heating needs of a house. Shading devices must be used over the windows to exclude unwanted solar radiation.

The challenge confronting passive solar design is that of storage and control of the heat gained in this manner in order to maintain suitable comfort standards within the building. The approach customarily taken is to utilize a large storage mass. This storage mass is either an integral part of the structure or is located inside the building. Storage of heat is commonly done with dense, interior materials that have a high heat-holding capacity such as masonry, adobe, concrete, stone, and water. These materials absorb surplus heat from the sunlight entering the windows and radiate it back into the room after dark. Some passive homes have walls up to a foot thick.

The building's thermal mass comprises massive structures such as thick floors or walls of stone, concrete, adobe, masonry, or even water. On many days the sun provides more energy than is needed to maintain the living and working space at a comfortable temperature. In this case the excess energy is used to warm up the massive floors, walls, etc. of the building. After sundown the building temperature will drop very slowly because of the extra heat released by the storage media. This thermal mass helps to smooth out the extremes of temperature within the building that would otherwise occur. It also helps maintain cooler indoor temperatures during the summer. The large, heavy structures absorb heat from the inside air during the daytime and release it at night.

The advantages of passive solar systems are so remarkable that this approach to solar heating (and cooling) has dominated the scene in recent years. In fact the period of time between 1975 and 1980 might be termed the "Passive Revolution." Some of these important advantages are tabulated below (this list is certainly not either exhaustive or unique):

1. **Simplicity of Design.** No complicated plumbing and control systems are needed. Active space heating systems (see Chapter 9) tend to be large, complex, and cumbersome.

2. **Increased Livability.** Occupants of passive buildings are usually extremely pleased with the living space. There appear to be many reasons for this. Passive structures tend to be more open to the environment, resulting in an increased daylight effect. Heating a house using warm walls or floors is generally more comfortable than when this is accomplished by heating air. Finally, passive buildings are quiet, peaceful places in which to live.

3. **Performance.** Passive structures have as good (or better) a performance record as their active solar counterparts. It has proven

simple to reliably obtain 50 to 80% of the annual space heating requirements in this fashion. Heating a house with a large surface at a low temperature is not only comfortable, but heat losses are reduced from the low-temperature collection and storage system, thereby increasing efficiency.

4. **Longevity.** There is no reason why the passive features of a building will not last for its useful lifetime.

5. **Economics.** In a real sense the passive features are part of the building and do not represent any additional cost.

The many attractive features of passive homes make them quite appealing to most people. Nevertheless, living in a passively heated home requires some adjustments on the part of the occupants. The temperature often fluctuates much more than is the case for a conventional home. This may mean wearing an extra sweater at times, paying extra attention to the backup heating system, and even living with some occasional unpleasant temperatures. You may also have to do a few extra daily chores such as opening and closing vents, dampers, insulating panels, and the like.

A. IMPORTANT DESIGN FACTORS

Designers are still grappling with the limitations of passive solar heating. Since direct sunlight cannot penetrate all parts of a typical house, an innovative architectural design is required. Some of these designs include having a shallow house on a long east-west axis, a "stacked" design like the Karen Terry house described later in this chapter, or skylights that bring sunlight to interior spaces. Other innovations include movable shading devices to control sunlight, movable insulation panels to reduce nighttime heat losses, and ventilation ports to either augment or reduce daytime heating by means of natural convection.

Buildings that are designed to take advantage of the sun's energy can greatly reduce the demands on heating and cooling systems. This can be especially important in reducing the cost of the active solar heating and cooling system needed. Some important factors to be considered in passive solar design are as follows:

1. A large expanse of south-facing glass is needed. Buildings should face north-south rather than east-west to reduce annual energy consumption. South-facing windows receive maximum winter gains and minimum summer gains. East and west walls should not have large exposures of glass.

2. The building should be very well insulated. Decreasing the building heat loss is the first step toward an energy efficient building.

3. Since passive solar energy collection cannot store up heat as well as active systems, there is not as much carry-over for cloudy days. A large mass is needed to provide sensible heat storage. This is often accomplished through the use of thick walls and floors. Another promising approach is through the use of a large surface area of rather thinner masonry. An outer layer of insulating material can be used advantageously. Thermal mass located in direct sunlight has roughly twice the effective thermal storage capacity as has mass located in a room that is open to the sun, but not directly exposed. In turn, the latter is twice as effective as mass located in a more remote region.

4. Interior open space can often be advantageous. This is an excellent way to increase the attractiveness of the building.

5. Glassed-in porches or sunspaces can be attached to a house in order to trap heat directly, exactly as in the case of windows. This arrangement permits a more exposed glazed area than for simple vertical windows. The greenhouse must be closed off at night to cut off heat losses when the sun is down.

6. Roof overhangs and window awnings can be designed to admit the low winter sun and block out high summer sun.

7. There are many different ways to eliminate heat loss through the windows at night or on cloudy days. For example, double-glazed windows or movable drapes can be used. The use of some type of movable insulation can reduce heat losses through the window, over that obtained with only single glazing, by as much as a factor of 5.

8. Even during the winter months, the gain of heat close to large expanses of window area may be too large. One way to compensate for this is to use a fan to give forced circulation of the air. Another method is to orient the window area to the east of south. This reduces the late afternoon solar gain, at which time the house is already at a comfortable temperature. Another method is to use exposed masonry surfaces on the interior of the space, facing the windows. The masonry (e.g., brick or concrete) absorbs heat and can then deliver it to the house later on (see the following discussion of the Trombe wall design).

9. The use of specular reflectors can significantly increase the total influx of solar radiation through a window. Furthermore, a reflector oriented slightly below the horizontal plane will enhance the

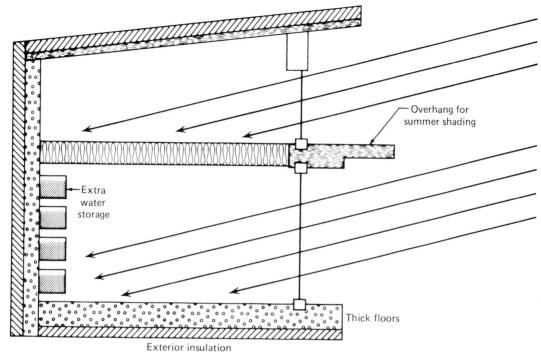

FIGURE 7-1 Example of a direct-gain system. During daytime in the winter the excess solar energy is absorbed by the large amount of thermal mass.

amount of light gathered in the winter while providing very little extra gain in the summer. Glare from a reflector below eye level can be a problem.

Passive systems vary considerably in their methods of collecting, storing, and distributing heat. Here the most important approaches are discussed. For orientation, we first comment briefly on each type, emphasizing the method of heat collection, its location with respect to the remainder of the building, and the arrangement for heat transfer.

In a direct-gain system the actual living space is warmed by sunlight entering through a south-facing window. This approach is illustrated in Figure 7-1. Excess heat is stored by thermal mass in the form of thick floors and walls. In systems utilizing a thermal-storage wall, the sun's energy is stored in a thick masonry or water wall located just behind the south-facing glazing. Roof ponds feature plastic bags filled with water and are located on top of the building. Both of these approaches utilize indirect gain, with the solar energy absorption taking place at the surface of the building. Some of the absorbed heat is used directly, whereas the remainder is stored for later use. In this case the residents interact only with the storage and distribution elements. A solar greenhouse is an example of an isolated gain system in which collection occurs at a location that is either thermally isolated or separated physically from the living area. Other examples of this approach involve a thermosiphoning air heater and the use of rockbeds. Heat can be stored at higher temperatures using this approach than is the case for the other passive system types. Finally, we mention the important passive approach using superinsulation; the double-shell house is included in this category.

B. DIRECT-GAIN SYSTEMS

The direct-gain system is perhaps the simplest approach to passive solar heating. A large south-facing expanse of glass is utilized to permit direct entry of the solar radiation into the living space. The building itself becomes a live-in solar collector. Heat storage is accomplished with the aid of thick floors and walls constructed of concrete, stone, masonry, or adobe. Additional storage using water drums is often included, as illustrated in Figure 7-1. The warm air of the living space facing the south window area will rise and can be vented along the ceiling to other rooms.

The nighttime operation of a direct-gain house is illustrated in Figure 7-2. Movable insulating panels are often used to cover the windows when energy is not being collected, thus minimizing the heat loss during this period. The energy absorbed by the walls and floors is given back to the living area during the night period. During the summer the overhangs prevent excessive sunshine from entering the large direct-gain window and clerestory. The rule of thumb that applies for the overhangs is that the windows should be fully shaded at a noon sun elevation of 78° (above the horizontal) and should be just shaded at the top at a noon sun elevation of 45°.

It is important to choose the correct amount of south-facing window for a direct-gain building. Clearly the exact amount depends on the fraction of space heating to be provided by the solar system and on the local climatic conditions. A rough guide is to use 1 ft² of south-facing glazing for each 4 ft² of floor area. It is often possible to orient the building with its long axis along the east-west direction; this increases the potential for solar gain along the south side.

The amount of thermal storage mass needed can be obtained from the following rule of thumb. Use at least 40 lb (18 kg) of water or 150 lb (68 kg) of rock or masonry for each square foot of south-facing glass. If the storage mass is not located in the direct sun or in a room exposed to

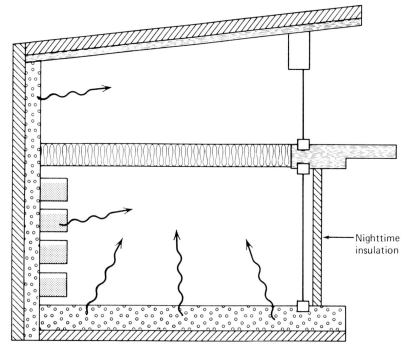

FIGURE 7-2   Direct-gain system during night hours. The solar energy absorbed by the thermal mass is returned to the living space.

the direct sun, the amount required is about four times larger in order to provide the same range of temperatures in the living space.

One of the earliest examples of a direct-gain system is the St. George's County Secondary School in Wallasey, England near Liverpool. It was designed by A. E. Morgan to house 300 students and to gain its entire heating needs from solar energy, lighting, and the students. A schematic drawing of this structure is shown in Figure 7-3. The basic design approach was to make the structure sufficiently massive so that thermal inertia would prevent rapid or extreme fluctuations in the internal air temperature. In addition, the exterior walls are insulated with 13.5 cm of polystyrene.

The building is a long, thin two-story structure running east and west. The construction is concrete, with the ground floor consisting of 10 cm of screen on 15 cm of concrete, the intermediate floor being a 22-cm slab of concrete, and the roof being 17 cm of thick concrete. The partition walls are 22 cm of plastered brickwork. On the south side, the external wall, shown in Figure 7-4, is mostly double-glazed and is 8.2 m tall and about 70 m long. The outside sheet of glass is clear and is

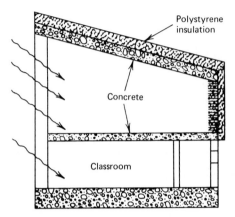

FIGURE 7-3 Cross-sectional view of the Wallasey annex to St. George's school. Sunlight enters through the large double-glazed south wall. Stable temperatures are maintained through the use of massive floors and walls and by the use of polystyrene insulation to help reduce building heat losses.

spaced about 0.61 m away from the inside sheet of *figured* glass—a type of glass that refracts the sun's rays in such a way that both the ceiling and the floor are irradiated in a uniform manner. Each classroom, however, is provided with two or three openable *view* windows,

FIGURE 7-4 Photograph of the Wallasey annex, showing the large south-facing expanse of glass. (Courtesy of R. McDonald.)

FIGURE 7-5 Picture of the passive solar home of Karen Terry in Santa Fe, New Mexico. (Courtesy of E. Mazria.)

which constitute areas of single glazing. Measurements of the thermal performance of the system indicate that the maximum fluctuations during winter days is about 2 to 3°C.

One way to overcome the problem arising because direct sunlight cannot penetrate all parts of a typical house is to use a *stacked* design. This approach was taken by architect David Wright in designing the Sante Fe solar house of Karen Terry shown in Figure 7-5. The house is oriented to receive the low winter sun, receiving and storing heat on all levels, as indicated in the sectional view shown in Figure 7-6.

The solar house is set deep into the hillside with earth berms piled up against the side and the back. The house measures 5.5 by 16 m with 37 m^2 of double-glazed glass in four levels. During the summer the glass is shaded by removable wooden louvers. Between the lower and middle levels is a wall consisting of eighteen 208-liter steel drums containing water and an anticorrosive additive. Similar drums are used to form the barrier between the middle and upper levels. At night, 5-cm polystyrene insulating shutters swing up to cover the glass, helping to hold in the heat that has been absorbed during the day. The normal temperature range on a typical winter day is from 15°C at sunrise to 24°C in the late afternoon. Backup heat is provided by a fireplace and a woodstove.

It is sometimes desirable to utilize a window area that rises above the adjoining roof. This window area is termed a clerestory. The same

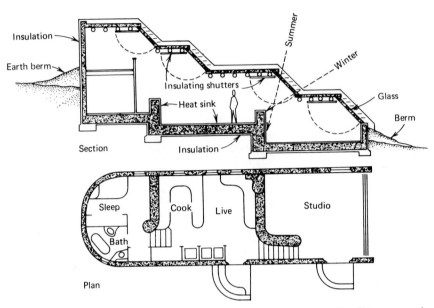

FIGURE 7-6 Plan and sectional views of the Terry solar house. The living space is arranged on three levels, while the fenestration is on four levels.

design principles outlined above apply. Movable insulation must be provided for times when the sun does not shine (if not, double glazing is then absolutely required). If a vertical window is used as a clerestory, then the roof just below can be used as a reflecting surface to increase the amount of winter sun obtained. Skylights can also be useful direct-gain devices, but in this case special care must be taken to provide shading in the summer months.

With direct-gain systems excessive solar gain in the fall and spring should be avoided. In some passive houses the rooms have become so warm during this time period that doors had to be opened. Glare and fading of furniture and rugs can also be a problem.

C. THERMAL STORAGE WALLS

Another novel passive solar approach uses a large, massive wall for thermal storage. This is exemplified by the Trombe-wall houses constructed in the solar community near Odeillo in the French Pyrenees by Felix Trombe and his collaborators. The design principle is illustrated in Figure 7-7 and features a massive south-facing, vertical concrete wall, which is painted black and covered with a sheet of glass. An airspace runs between the concrete and the glass; the chimney effect causes the heated air to rise. During the day openings at the bottom and

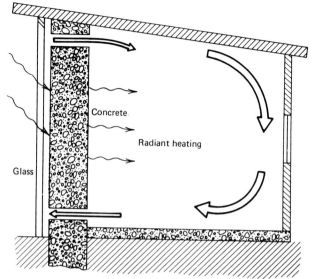

FIGURE 7-7 Trombe wall-type of passive solar house. The house faces south with the air heated by the incoming solar radiation, which is absorbed by the black face of the wall.

top of the wall allow the cold air along the floor to enter the air space. Similar openings at the top provide an opportunity for the warm air to reenter the room behind the wall. The air then circulates through the room, warming the walls and the occupants with a portion of the energy collected.

The remaining energy is transmitted by conduction through the wall, resulting in a rise in temperature at the wall's interior surface. Energy is then spread through the living space by conduction, convection, and radiation.

Data taken from a Trombe-wall building constructed in 1967, which had walls 0.60 m thick, indicated that one-third of the solar radiation incident on the south wall during the winter months was transferred into the house. About 70% of the space heating load was provided by solar energy. Of this amount, about 20% was transported into the living space by convection, and the remaining 50% came by conduction through the wall. In the summer, the glazed concrete wall was shaded from the sun by the overhanging roof.

The Trombe or thermal-storage wall is an indirect-gain system that works by absorbing and storing heat in a mass wall. This stored heat is then transferred to the living space of the building. The entire exterior face of the storage wall is painted black and is covered by one or two layers of glazing, located several inches in front of the wall. In what follows the basic elements of the Trombe wall will be studied: glazing, air space between glazing, storage wall, night insulation, vents, and roof overhang.

A Trombe wall with vents is illustrated in Figure 7-7. With an unvented wall, heat is stored in the wall during the day, is conducted through the wall, and is then transferred to the living space. A vented Trombe wall facilitates the daytime transfer of heat to the house. Cool air moves through the lower vents into the space between the wall and the glazing while warm air passes through the upper vents into the building. About 30% of the total heat collected by the system is convectively transferred to the building in this way. With a vented system, adequate thermal storage within the building must be provided to prevent overheating during the day.

The glazing must have the usual properties: resistance to ultraviolet ray deterioration, good thermal stability, ability to stand harsh weather, strength, and low cost. Double glazing is usually recommended. This is dramatically illustrated in Table 7-1, which compares single and double glazing for two different thicknesses for the storage wall. The outer glazing should be durable as it is exposed to both the weather and ultraviolet radiation. The inner glazing should be able to withstand high temperatures. An unvented wall can reach as high as 200°F on the wall's dark outside face while a vented wall can reach 160°F.

As shown in Table 7-1, placing insulation over the glazing at night can also improve the system performance. Perhaps the most practical and economical solution is to use either single glazing plus night insulation or double glazing. One can also use a selective surface (see Chapter 8) attached to the masonry wall. This has the same effect as using double glazing with R-9 night insulation (this is roughly equivalent to 7 cm of fiberglass insulation). The recommended spacing between the wall and the inner glazing is 7 to 15 cm. A larger spacing

TABLE 7-1
Effect of Glazing and Insulation on the Performance of a Thermal Storage Wall[a]

GLAZING TYPE	WALL THICKNESS	
	15 cm (6 in.)	30 cm (12 in.)
Single	27	32
Double	71	79
Single and night insulation	81	89
Double and night insulation	90	96

[a] All numbers in the table represent the percentages of the total building heat contributed by a water thermal-storage wall system. Data source is "Passive Solar Buildings: A Compilation of Data and Results," R. D. Stromberg and S. O. Woodall, Sandia Laboratories Report, 1978.

may result in excessive heat losses from the sides and top of the framing structure.

The important elements of the storage wall itself are its size, conductivity, and thickness. The effect of varying the conductivity is shown in Figure 7-8. If the conductivity is low, as is the case for wood (0.10), heat cannot easily pass through the material. The higher the conductivity, the more readily heat can move from the outside wall to the inside wall. This allows a thicker wall to be more effective than is the case for a low-conductivity material. While the conductivity of water is 0.35 (all units are Btu/ft-hr-°F), in effect it is almost infinite because of the rapid transfer of heat by convection. With a water wall usable heat reaches the living space from the inside wall during the day. The disadvantage is that cool-off in the evening will be quicker with a water wall than with a concrete storage wall (conductivity = 1.0).

From Figure 7-8 one would infer that the optimum thickness of a concrete storage wall is between 9 and 16 in. This optimization is in terms of the efficiency of the Trombe wall in which the annual solar heating was calculated. However, in terms of comfort, a somewhat

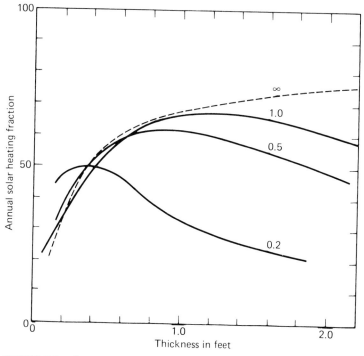

FIGURE 7-8 The annual solar heating contribution in percent is plotted as a function of wall thickness for several different values of wall conductivity.

thicker wall may be better. This is illustrated in Figure 7-9, which shows that a thicker wall will have smaller variations in the temperature of the inside wall. Clearly, a 6-in. concrete wall allows much too wide a range of inside temperatures; a 24-in. wall provides much more comfort with only a small sacrifice in the annual solar heating contribution.

The area of thermal storage wall to use in a building is a function of both the local climatic conditions and of the building's insulating properties. It is common practice to express the latter quantity in terms

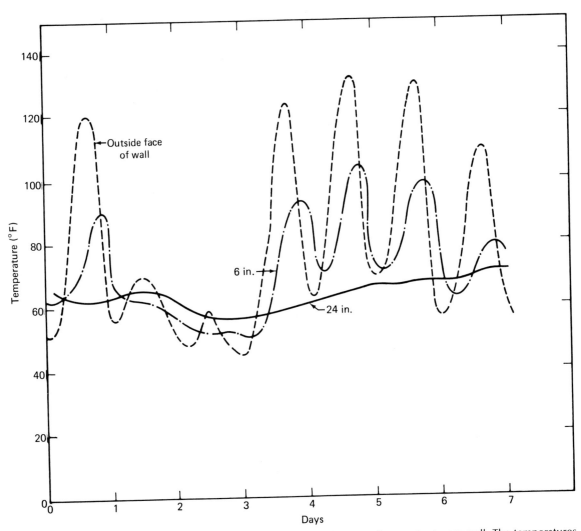

FIGURE 7-9 This illustrates the performance of two different thicknesses of concrete storage wall. The temperatures for the 6- and 24-in.-thick walls are at the inner face of the wall.

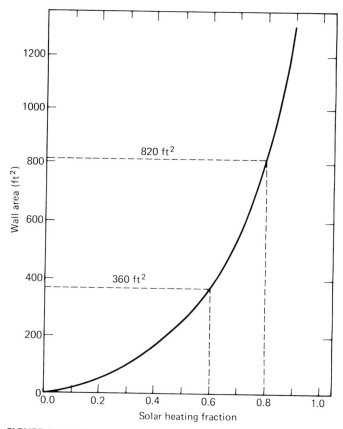

FIGURE 7-10 The area of thermal storage wall required for Eugene, Oregon is plotted as a function of the fraction of the annual heating load contributed by the solar system.

of Btu's per degree-day. Then, knowing the number of degree-days, the annual heating load can be calculated. Once the total heating load of the building is determined, it is then necessary to calculate the size of storage wall needed to provide the desired amount of solar heating. The results for Eugene, Oregon are depicted in Figure 7-10 for the case of a building with a heating load of 12,000 Btu/degree-day. The calculations were based on tables derived by Balcomb and McFarland of the Los Alamos Scientific Laboratory. The most important result to be gleaned from Figure 7-10 is that the wall area required increases rapidly as the fractional load carried by the solar system rises. As shown, a 33% increase in the solar fraction from 0.60 to 0.80 requires that the wall area increase by a factor of 2.28 (an increase of 128%).

A fixed overhang from the south roof provides shading in the summer. The length of the overhang should be sufficient to provide

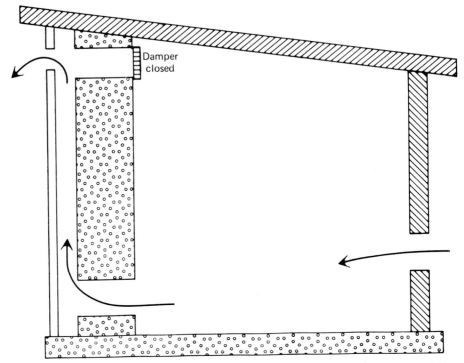

Damper
closed

FIGURE 7-11　Schematic drawing of a Trombe wall system utilizing vents for summer cooling.

nearly complete shading for the noon sun in late June. The proper operation of vents will also aid in providing summer cooling. This is shown in Figure 7-11; the upper wall vent should be closed and the lower wall and upper glazing vents should be open. By opening windows on the north side of the building, a convection current is set up that pulls cooler air through the north windows, across the living space, and out through the vent in top of the glazing.

D.　OTHER PASSIVE IDEAS

There are a large number of passive approaches. We have described direct-gain and Trombe wall systems in considerable detail because they allowed us to illustrate many important concepts. In the following discussion some of the other important passive ideas that have found widespread usage are briefly studied. The reader needing more detailed information can find it in one or more of the bibliographical items listed at the end of the chapter.

D.1 Sunspaces

A sunspace, or solar greenhouse, is usually attached to a south-facing wall, although buildings sometimes incorporate sunspaces on east or west walls provided some southern exposure can be gained. The thing that differentiates a solar greenhouse from any other is that all of its energy is derived from the sun and that provision is often made to store some of the energy captured during the day for future use. This prevents overheating during the day and smooths out the temperature swing during the night.

An example of an attached sunspace is illustrated in Figure 7-12. This additional sunspace can make a positive change in the winter living conditions. It provides the building's occupants with an outdoor atmosphere while maintaining a moderate climate. In this area plants and vegetables can be grown. The sunspace is usually allowed to "float" in temperature on winter evenings. This means that it is allowed to drop in temperature with no heat input from the house. Homes can usually be retrofitted with an attached sunspace because it is separated from the rest of the house by a wall.

As shown in Figure 7-12, heat from the sunspace can enter the house through doors, windows, or vents. Excess heat collected on good solar days can be stored in the thermal mass of the house by circulating air

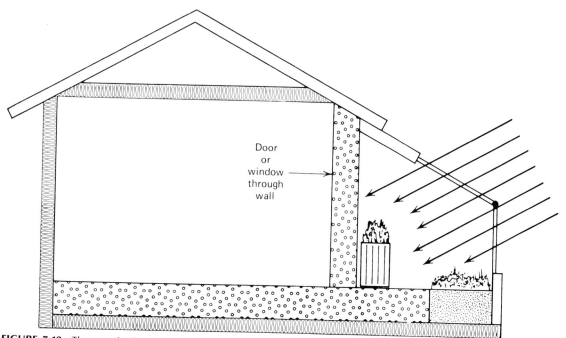

FIGURE 7-12 The attached sunspace can provide some extra heat for the house while providing an attractive sunspace for increased livability.

Door
or
window
through
wall

from the greenhouse throughout the house, either by natural convection or with the aid of a fan. The greenhouse itself is protected against severe temperature fluctuations by having sufficient storage mass. The rule of thumb is to use $1\frac{1}{2}$ to 3 ft^3 of rock storage for each square foot of south-facing glass. The connecting wall between the greenhouse and the main structure can even be a Trombe wall. In this case the house need not be shut off from the greenhouse at night. Otherwise, closing off the house to the greenhouse during severe winter nights is recommended in order to prevent heat loss from the main portion of the building.

D.2 Superinsulated Houses

A superinsulated building is one that has such small heat losses that very little auxiliary heat is needed to maintain a comfortable temperature. Most of the building's heat losses are provided for by normal occupancy. This means that the major sources of energy are human bodies, windows, and electrical appliances. In point of fact, the solar considerations involved are quite small. We include this approach, however, because reducing the heat losses from a building is a prime consideration for all passive solar approaches. The superinsulated house merely carries conservation measures to the extreme, so that only a small amount of direct gain is needed.

Proponents of superinsulated houses point out that building heat losses are so low that no backup heating is needed in midwinter, even in extremely cold climates. There are a number of essential features that make it possible to have a structure with such a low heat loss. All external surfaces including the basement are insulated with 20 to 30 cm of fiberglass or the insulating equivalent. A very airtight construction is utilized involving a carefully installed vapor barrier. Windows are usually triple-glazed. Window area is moderate, never exceeding 10% of the floor area and it is concentrated on the southern exposure of the building.

Perhaps the best example of this design approach is the Saskatchewan Conservation House, built in Regina, Saskatchewan by the Canadian government to demonstrate that conservation techniques can work even in a very cold climate. A drawing of this house is shown in Figure 7-13. The Saskatchewan house is superinsulated, with fiberglass and cellulose in the walls providing R-40 insulation (about 40 cm of fiberglass insulation), a ceiling insulation of R-60, and a crawlspace under R-36 floor insulation. Windows are double-glazed downstairs, triple-glazed upstairs, and have insulated shutters. The house inside is virtually airtight and free of drafts because of a vapor barrier three times thicker than standard plastic sheeting. All electrical outlets were isolated in airtight plastic enclosures. The front entry is equipped with three doors: an outside storm door, then a heavy solid core door, and finally another tight door located at the other side of a small vestibule.

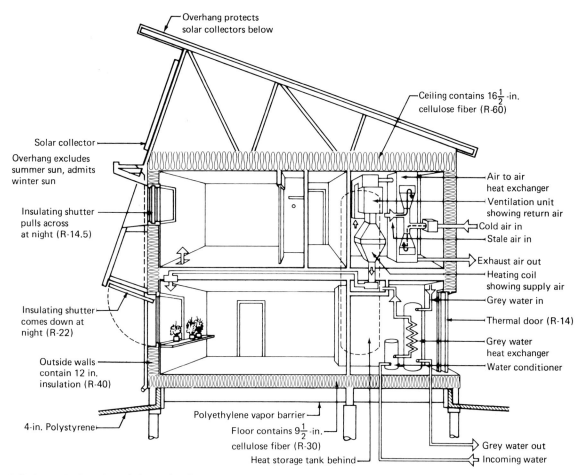

Overhang protects solar collectors below

Ceiling contains $16\frac{1}{2}$-in. cellulose fiber (R-60)

Solar collector

Overhang excludes summer sun, admits winter sun

Insulating shutter pulls across at night (R-14.5)

Air to air heat exchanger

Ventilation unit showing return air

Cold air in

Stale air in

Exhaust air out

Heating coil showing supply air

Grey water in

Insulating shutter comes down at night (R-22)

Thermal door (R-14)

Grey water heat exchanger

Water conditioner

Outside walls contain 12 in. insulation (R-40)

4-in. Polystyrene

Polyethylene vapor barrier

Floor contains $9\frac{1}{2}$-in. cellulose fiber (R-30)

Heat storage tank behind

Grey water out

Incoming water

FIGURE 7-13   Drawing of the Saskatchewan superinsulated house.

An added conservation feature is a grey water tank that cuts 30% off the energy needed for hot water heating. The overall thermal mass of the building is rather small.

The thermal performance of the Saskatchewan house has been exceptional. The gross heat loss per year is 43,500 MJ (million joules). After deducting 14,200 MJ for passive solar gain, 4600 MJ for human gain, and 19,600 MJ from electrical appliances, the net space-heating requirement is only 5100 MJ (1417 kWh). This is less than 4% of the space heating requirement for an average Regina house.

One of the key conservation measures involves stopping infiltration, which is often the largest single contributor to energy loss from the building. This can lead to poor inside air quality unless steps are taken. The solution is to use an air-to-air heat exchanger. This device uses a small fan to pull fresh outside air into the house while at the same time forcing out the stale, inside air. The two air currents are made to flow past each other in opposite directions so that most of the heat in the outgoing air is transferred to the incoming air. This keeps the inside air pleasant and fresh while venting very little heat to the outside atmosphere.

D.3 Double-Envelope Houses

This approach to passive solar heating has evoked a considerable amount of controversy. The controversy is not over the system performance, as double-shell houses appear to work reasonably well. Instead the question is why do they work? We will come back to this aspect later. First, let us define exactly what is meant by a double-envelope house. The customary definition is that it is a house combining a south-facing greenhouse with a double-shell construction for the main building structure. The latter results in a continuous air plenum surrounding the house. Some houses built in this manner appear to have an auxiliary heating requirement of only one cord of wood for a 2000 ft^2 house in a 7000 degree-day climate [see Table 9-1].

The basic construction features of a double-envelope house are illustrated in Figure 7-14. The arrows on the figure show the proposed circulation of air according to the original thinking on the subject. The air space between all walls is at least 15 cm, with that in the attic and

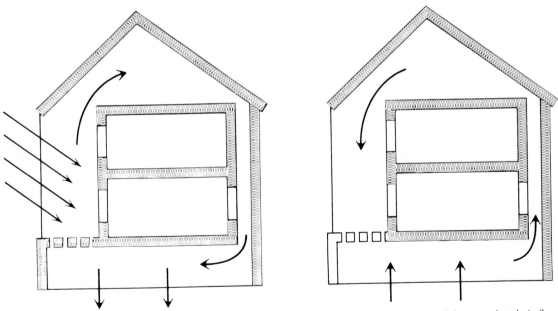

FIGURE 7-14 Schematic illustration of double-envelope house. See text for a discussion of the postulated air flow.

the crawl space under the floor often considerably larger. Both the inner and outer walls are well insulated. In the conventional design illustrated in the figure the air in the crawl space beneath the floor is in contact with the floor. Additional thermal storage mass can be placed beneath the floor and the crawl space insulated from the ground as an alternative.

It was originally thought that the double-envelope house worked because of the air flow shown in Figure 7-14. During the day the air warmed by the sun in the greenhouse would be displaced upward as cool air moved in from the crawl space resulting in a clockwise circulation of air. Some heat would be stored in the crawl space. At night the air flow would reverse, according to this theory, because of the large heat loss from the greenhouse.

At first glance this hypothesis appears attractive. But, under closer inspection, it is found to be unsound. The energy available simply isn't large enough to account for a significant circulation of air. Low air-flow rates would not allow for the transfer of significant amounts of heat into storage.

It now appears that the basic reason for the success of the double-shell scheme is that the temperature of the air between the shells stabilizes at a value intermediate to the interior and exterior temperatures, but closest to the warmer interior value. The heat to maintain the temperature of the loop comes either from the greenhouse on warm days, from the earth on cool days, or from both sources at times. Viewed in this way, the double envelope's performance is easily understood. The heat loss through the inner shell (at times even a gain occurs) is greatly reduced over that which prevails with a normal one-wall structure. The double-envelope building has attained the same goal as the superinsulated house, but with an entirely different approach.

Proponents of this approach can point to numerous advantages. A well-designed double-shell structure requires no appreciable auxiliary heat. The airspace between the walls allows increased glazing for the interior walls without a significant increase in heat losses. On the other hand, most architects and builders agree that the double-envelope design is slightly (perhaps 5 to 10%) more costly than a comparable conventional home. A potential fire problem exists in the air plenum and methods to protect against this incur an additional cost penalty. Finally, the north windows of some envelope buildings are fogged all winter. Clearly, much remains to be learned about this intriguing and controversial structure. And it must be stressed that obtaining a firm understanding of the physical mechanisms responsible for the building's performance is essential. Otherwise, real progress in future double-envelope designs will be impossible.

D.4   Rockbeds

An important aspect of any passive building is the storage of excess solar heat. With a Trombe wall this is accomplished at the absorber, but in most systems this is not the case. A direct-gain house, for example, with 250 ft^2 of south-facing glass needs roughly 750 ft^2 of 1-ft-thick concrete placed directly into the sun to store the excess heat; often, however, this is not easily accomplished and the designer needs to look for an alternative. A favorite choice for this application when the excess energy is in the form of hot air is the fan-assisted rockbed (assuming the rockbed lies below the source of hot air).

This rockbed approach is particularly attractive when the hot air comes from an attached sunspace. Removing heat from the sunspace then reduces the air temperature and makes the sunspace more livable. The stored heat can be retrieved at a later time and used to heat the main living area. Using rockbed storage in this way also helps to correct the tendency for inbalance in temperature within the passive structure due to the presence of most of the heat-absorbing element on the south end of the building. A major disadvantage of the rockbed approach is that oversized grills and ductwork must be used in order to move relatively large volumes of air that are not too much higher in temperature than that of the normal living space.

Rockbeds can be operated in a number of ways. One particularly simple approach is to force warm air through the bed of rocks and thence into the building. In this way a temperature wave moves through the bed, with a fan being kept on most of the time. This strategy results in cool air being forced into the house as the rockbed warms up. More often a two-way active-discharge rockbed is used. Heat is collected by forcing air through the rockbed in one direction and is distributed by forcing air through the bed in the reverse direction. In this way, the heat in the rockbed is stratified and the heating efficiency is improved as the warm air is returned to the house first. This approach requires more expensive controls and dampers. Experience indicates that rockbeds work only marginally when warm air is not forced in.

A good approach is to use a radiant-release system in which heat is removed from the rockbed by the physical processes of radiation and natural convection. A schematic diagram of this scheme is shown in Figure 7-15. In this scheme the rockbed is connected directly to the space to be heated. While it is customary to place the rockbed under the floor of the building, as indicated in the figure, it is also possible to place it behind a wall. Other advantages of this approach are that controls are simple, operating costs are low, and heat distribution is no problem.

The approach outlined in Figure 7-15 works because there is an excess of heat in the attached greenhouse. This results in higher temperatures in the sunspace than in the normal living area. In this case the fan-assisted rockbed is effective. Care must be taken with the

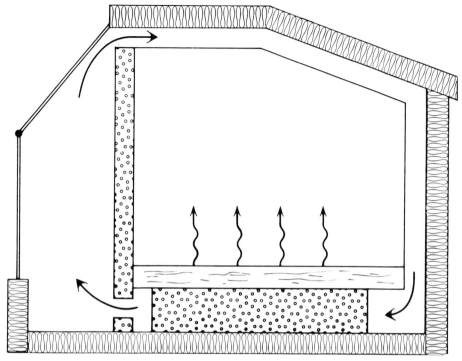

FIGURE 7-15 Radiant-release rockbed operating in conjunction with an attached greenhouse. The rockbed is heated by warm air from the sunspace during the day and returns the stored heat to the living space at night by convection and radiation.

design to ensure that this condition prevails. An important rule in radiant-release rockbed design is to maximize the surface area between the bin and the living space. Also, the exterior surface of the rockbed should be sealed and insulated.

Moisture associated with condensation of humid air in the rockbed could lead to odors or harmful fungi. Usually, this moisture will evaporate when the rockbed temperature rises. Moisture is almost always a problem when a rockbed is used for summer cooling. Another problem that has been reported concerns the settling of the rocks in the bed. This can cause "channeling," a short circuit of the airflow over the top of the rocks. A layer of sand on top of the rockbed can be used to prevent this.

D.5   Roof Ponds

Another simple and elegant passive solar system uses roof ponds. In this system water is contained in a clear plastic bag and placed on a black metal roof. This idea was designed and patented by Harold Hay, who first got the idea while in India on a technical aid mission for the

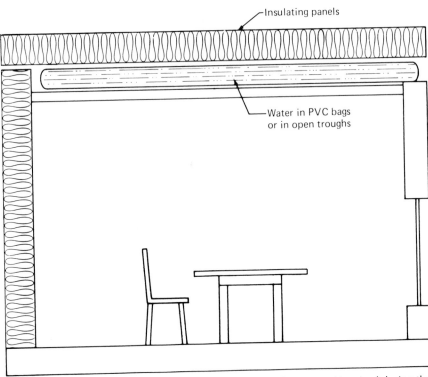

FIGURE 7-16 Schematic of a roof pond system. The insulating panel is raised during the day and closed at night for winter heating operation. For summer cooling the cycle is reversed so that the insulating panel is raised during the night period.

U.S. government. He noticed that many people lived in rusty, sheet-metal shacks, which were hot in the day and cold at night. Hay's innovation was to remove the insulation from the roof on winter days so that the roof would get hot. Replacing the insulation at night would trap this heat, allowing it to warm the shack throughout the night. During the summer a reverse procedure would be followed, letting the house cool at night and replacing the insulation in the daytime to keep out the heat.

Over the years this idea was refined and in 1966 and 1967, Hay, in collaboration with John Yellott of the University of Arizona, construct-ed a 3- by 3.7-m building using water basins as the actual roofing material. During the summer, a slab of foam insulation was rolled back at night, and the water became cold through night sky evaporation. Since the water reservoir sat directly on a metal ceiling it absorbed heat from the room and kept the building air-conditioned all day. During the winter the movable insulation was rolled back in the daytime in order to collect heat as shown in Figure 7-16. It then radiated enough heat into the house through the ceiling at night to keep the room comfortable. Tests over an 18-month period seem to have justified the designer's predictions. Hay notes that when the temperature outside was 36°C he had to wear a sweater and a coat inside to keep warm.

This approach should be effective in dry climatic regions. An improved version has been tested in Atascadero, California. The exposed roof area was about 102 m², approximately equal to the floor area. The average water depth of the roof ponds was about 20 cm, giving a total of 22,700 liters of water. The system was operated both with and without glazing. The glazing, an inflatable clear plastic cover, proved necessary in order to maintain the indoor temperature during the winter months. This solar system supplied 100% of the house's heating and cooling requirements. The indoor temperature was maintained between 14 and 23°C. The weight of the roof pond apparently created no major problems. Problems with roof leaks developed, but were corrected. In areas of extreme cold during the winter, the roof pond systems need additional protection beyond the insulating ponds to prevent freezing. Often the roof pond water is contained in polyvinyl chloride (PVC) bags. The ultraviolet rays in sunlight will cause the PVC bags to deteriorate seriously over a period of years, requiring that the bags be replaced.

REFERENCES

1. *Sunset Homeowner's Guide to Solar Heating* (Menlo Park, CA.: Lane Publishing Co., 1978).

   Contains an excellent and readable discussion of many features of solar-heating systems. This is a very nicely illustrated book.

2. E. Mazria, *The Passive Solar Energy Book* (Emmaus, PA: Rodale Press, 1979).

   A very readable and complete book on passive solar applications. The illustrations are excellent and novel.

3. W. A. Shurcliff, *Superinsulated Houses and Double Envelope Houses: A Survey of Principles and Practice* (Andover, MA.: Brick House, 1981).

   Shurcliff is the author of several good books on solar applications. This volume presents several illustrations of these two new passive approaches. The strengths and weaknesses of each are discussed.

4. "Passive Solar Design Handbook: *Passive Solar Design Concepts*, Vol. 1, and *Passive Solar Design Analysis*, Vols. 2 and 3." (U.S. Department of Energy, Washington, D.C., 1980).

   A large collection of information and data based primarily on the work of the Los Alamos solar energy group.

5. *Proceedings of the Annual Passive Solar Conferences* (Denver, CO.: American Solar Energy Society).

The American section of the International Solar Energy Society has sponsored a number of conferences on passive solar applications (seven through 1982). The proceedings of these conferences contain a wealth of useful information on all aspects of the subject.

FILMS

1. *A Place to Live*, 16 mm, color, 24 min. 1978. Produced and distributed by Fred James (Lumen-Bel, Inc., New York).
   Describes passive solar homes in Maine. This film describes how to effectively use those resources that are naturally available.

2. *The Solar Frontier*, 16 mm, color, 24 min. 1978. Produced by Mellenco Films. Distributed by Bullfrog Films, Oley, Pa.
   This film examines three Canadian solar homes, one active, one passive, and one hybrid. An interesting aspect of this film is that the architects give reasons for the various design aspects.

3. *The Solar Promise*, 16 mm, color, 28 min. 1979. Produced by Henry Mayer. Distributed by Bullfrog Films, Oley, Pa.
   An award-winning film that is outstanding technically and pictorally. This is an excellent film to use as an introduction to solar energy.

4. *A Building in the Sun*, 16 mm, color, 20 min. 1980. Produced by Gary W. Griffin. Distributed by Bullfrog Films, Oley, Pa.
   The Plant Science Building of the Cary Arboretum of the New York Botanical Garden is built from the ground up in this film. This is not a technical film. The audience can relax and enjoy the mood conveyed by the visuals and the accompanying music.

QUESTIONS

1. Why is a large amount of thermal mass necessary for a good passive solar building?

2. Name three important design features for passive solar heating.

3. What is the purpose of the polystyrene on the Wallasey solar building?

4. Explain why the performance of a Trombe wall, constructed of water-filled tanks, does not decrease as the wall thickness increases beyond 50 cm, in sharp contrast with the performance of a masonry wall, which becomes less effective as the wall thickness increases beyond 50 cm.

5. One of the following is not a key element in the design of a passive solar house:

   (a) Large direct solar gain.
   (b) Roof sloped at a tilt angle of latitude plus $10°$.
   (c) Large mass to provide for thermal heat storage.
   (d) Good insulation to minimize heat losses.

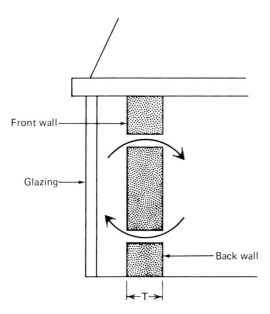

Front wall

Glazing

Back wall

|←T→|

The above drawing shows most of the basic elements of a thermal-storage wall system. The next 13 questions refer to this system.

6. For a concrete or masonry wall what should the wall thickness T be to maximize the amount of solar energy collected?

7. The answer to question 6 is considerably smaller than almost all practical Trombe-wall designs. Why does one use a thicker wall, such that the actual solar energy collected over the winter heating season is less than the maximum possible.

8. What is the principal disadvantage of single glazing in front of the wall as compared to either double glazing or single glazing with night insulation?

9. Should one close the vents at night? Explain.

10. What is the color of the front wall? Explain.

11. How is heat transported from the front wall to the interior of the house? Assume a concrete storage wall.

12. Suppose the outside (ambient) temperature is 60°F and the glass temperature is 90°F. Is the front wall temperature closest to 80°F, 120°F, or 180°F?

13. If the vents were accidentally left closed during the day, could heat still be transported to the interior of the house? Explain.

14. What change would you suggest to improve summer cooling performance?

15. In Eugene, Oregon a house with a heat load of 12,000 Btu/degree-day can obtain a solar load fraction of 0.60 with a south-facing wall area of 365 ft². To obtain a solar load fraction of 0.80, an area of about 850 ft² is needed. For an economical Trombe-wall design, which one of these two options would you choose? Explain your choice?

16. Which has the greatest heat loss, a system with single glazing or one with single glazing plus night insulation?

17. Suppose the temperature at the front face of a water Trombe-wall is 95°F. What is the probable temperature at the back face of the Trombe wall?

18. Let us compare plastic and glass for the window material. Indicate which of the two is the largest for each item.

ITEM	GLASS	PLASTIC
(a) Cost per square foot	_____	_____
(b) Weight per square foot	_____	_____
(c) Lifetime as a useful glazing material	_____	_____

19. Explain how summer cooling is obtained with the roof-pond approach.

20. List four distinct types of passive solar buildings.

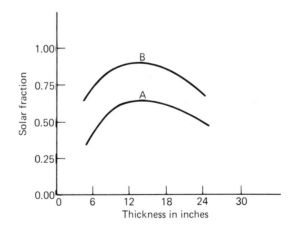

The above figure shows the performance of a thermal-storage wall system. The following five questions refer to this figure.

21. Suppose that curves A and B represent single-glazed Trombe wall systems. Which curve represents a system with night insulation?

22. Suppose that curves A and B represent single-glazed and double-glazed Trombe wall systems. Which is the double-glazed one?

23. What is meant by solar fraction?

24. What is the maximum solar load fraction attainable?

25. In practice, one does not build a system to obtain the maximum solar load fraction. Explain.

26. What is the distinctive feature of the Saskatchewan solar house that makes it a good passive solar design?

27. Name three common materials used in the wall of a thermal-storage-wall passive solar system.

28. A solar energy consultant advises a family not to put a rug on the floor in a particular room of their passive solar house. Explain what must be the situation and the consultant's reasoning.

8 capturing solar energy

Now that we understand the physical nature of solar radiation, it is time to consider how the incoming solar energy is captured and converted to useful purposes. From a very general point of view, the energy of the incident radiation is converted into two forms. The most common transformation is to heat up a gas, liquid, or solid and then transfer heat from this hot source to the place of need on demand. Commonly this thermal heat is used for water or space heating, cooling, or for process heat. An example of the latter application would be the use of solar-heated water to wash cans or bottles. The other general end use is the generation of electricity, either by thermal means or directly using some type of solar cell. The last two chapters are devoted to electrical generation schemes. For the present, we examine a few of the basic concepts about how the incident solar energy is concentrated to the point where high temperatures are achieved and electricity can be generated by thermal processes. First, however, we consider the simpler approach using flat-plate collectors where no concentration is necessary.

For solar energy applications that involve the heating of hot water and space heating, extremely high temperatures are not needed. A liquid (usually treated water) or air heated to 110 to 140°F will suffice. For these low temperature applications the simple, nonconcentrating flat-plate collector can be utilized. With this type of solar energy collection there is no need for the more complicated and expensive steering mechanisms that are required when concentrating devices such as lenses or parabolic reflectors are used. Another important advantage is that heat losses, which rise with temperature, are minimized. Finally, the flat-plate collector will collect solar energy from the diffuse component of the solar intensity, in contrast with concentrating devices which collect only the direct component for the most part. For these reasons, solar energy systems for water heating usually utilize some type of flat-plate collectors. Space heating with solar energy is sometimes done with flat-plate collectors, but more often the whole building is designed to be the collector and storage system. This approach is called passive solar and all of Chapter 7 was devoted to this important topic. The key principles of operation of a flat-plate collector can easily be taken over to the discussion of passive solar heating and cooling.

Devices for collecting and utilizing solar energy can be classified into four broad categories. Flat-plate collectors, generally used for space and water heating, are the most common example of this approach. These devices, which also include solar ponds and stills, accept and utilize the incident solar radiation without any concentration of the energy density.

The other three general categories utilize some degree of concentration of the incident solar intensity. This is achieved through the use of reflecting or refracting elements oriented in such a way as to concentrate the sun's rays onto the receiving portion of the solar collector. The first of these overall classes increases the solar energy density by a factor of 2 to 10 and requires only occasional adjustment to follow the sun. Straight reflectors and the compound parabolic concentrators are examples of this group.

For higher-concentration ratios, collectors that track the sun continuously are needed. These can be divided into two different classes. In the first, the collector is symmetrical about a plane and the sun is tracked by rotation about a single axis. Concentration ratios typically vary from 10 to 100. An example of this type is the parabolic trough concentrator. The highest degree of concentration is obtained by devices that possess symmetry about a line. These types of collectors require tracking motions about two different axes, much like that required for the equatorial mounts of telescopes. An example of this approach is the solar tower facility at Barstow, California.

A.   FLAT-PLATE COLLECTOR ELEMENTS

An exploded view of a flat-plate collector using a liquid as the heat-exchange fluid is given in Figure 8-1. Most common flat-plate collectors consist of the following key components:

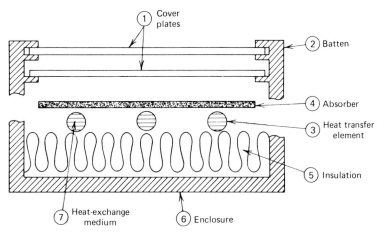

FIGURE 8-1   Exploded view of a flat-plate solar collector that utilizes tubes to transport the heat-exchange medium.

1.	Cover plate	This usually consists of one or two layers of glass or plastic.
2.	Batten (or cap strip)	Serves to hold down the cover plate (or plates) and provide a weather-tight seal.
3.	Heat transfer element	In the case of a liquid heat-exchanger fluid, tubes are often attached to the absorber plate. This facilitates the transfer of thermal energy from the absorber plate to the heat transfer medium. Tubes, roll-band plates, open channel flow, corrugated sheets, and finned tubes are some of many different types used.
4.	Heat-ex-change medium	Usually water or air.
5.	Absorber	This is usually a metallic plate coated with a black paint or black metallic oxide to improve the absorption of solar radiation.
6.	Insulation	Sufficient insulation is employed to reduce heat loss through the back and sides of the collector to a small fraction of the total collector heat loss.
7.	Enclosure	a container for all of the above components. The assembly is usually weatherproofed. Preventing dust, wind, and water from coming in contact with the absorbed plate, tubes, and insulation is essential for optimum performance.

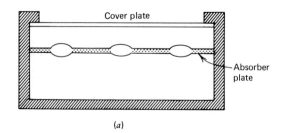

(a)

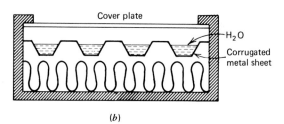

(b)

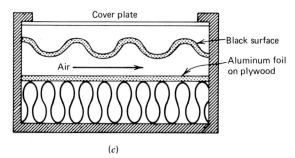

(c)

FIGURE 8-2   Examples of several styles of flat-plate solar collectors. (a) Tube-In-Sheet. (b) Thomason "trickle" collector. (c) Air collector.

Figure 8-2 shows schematic drawings of three other types of flat-plate collectors. The bonded sheet design, in which the tubes are integral with the absorber sheet, thereby guaranteeing a good thermal connection between the plate and tubes, is shown in Figure 8-2a. These sheets are manufactured using copper. Aluminum has a higher specific heat but a serious difficulty with aluminum is that pure water cannot be used as the heat-exchange medium because of corrosion problems; in this case a fluid containing corrosion-inhibiting agents must be used. Figure 8-2b shows the Thomason trickle-type corrugated channel collector, which is used with steeply pitched, south-facing roofs. One type of air collector is shown in Figure 8-2c. In this case, contact is made directly between the moving mass of air and the absorber plate.

An extended surface area is used to overcome the low values of the heat transfer coefficient between the metal and air (or glass and air).

B. COLLECTOR DESIGN

Glass has been the principal material used to glaze solar collectors because of its high transmittance (about 95% for high-quality glass of low-iron content) and almost complete opacity to the infrared radiation emitted by the hot absorber plate. The transmittance of a sheet of glass varies with the angle of incidence of the light on it, primarily because of the rapid increase of the reflected intensity at large angles. This is illustrated in Figure 8-3. For angles of incidence less than 50°, common glass reflects about 8% of the incident radiation with absorption accounting for another 4 to 6%. There are available techniques for making glass with a total transmissivity in excess of 95% for near normal angles of incidence. One possibility involves etching the surface in order to provide a gradation in density; this has the effect of

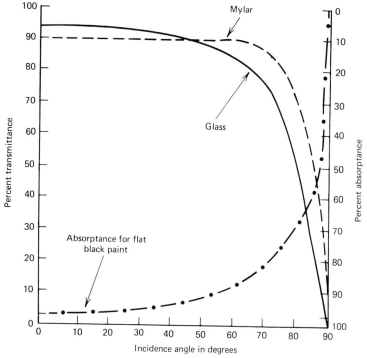

FIGURE 8-3 Transmission of light through glass and mylar as a function of the angle of incidence. The angle is measured from the normal to the glass or plastic surface. Also shown is the absorptance of flat-black paint.

reducing the reflected intensity. The same result can be obtained with dimpled glass, much like a shower stall door. A decrease in reflectivity can also be achieved with a thin coating of a material other than glass. Unfortunately, low-reflectivity glass is rather expensive.

The amount of absorption in the glass can be reduced by lowering the iron content. Detailed calculations with complete solar systems show that the decrease in system performance is considerably less than the reduction in transmissivity; a reduction of 1% in the absorption in the glass will result in only a $\frac{1}{2}$% increase in the useful energy obtained from the system. This is because the energy absorbed in the glass is not all lost from the collector, since this absorption raises the temperature of the glass, thereby reducing the convective and radiative losses from the absorber surfaces.

Plastic films have a high transmittance for the solar spectrum and have often been used as cover plates. Their transmittance curve, shown in Figure 8-3 for mylar, is often better than that of glass. Some disadvantages include a reduced opacity to long-wavelength radiation, a limited ability to withstand higher temperatures and scratching, and most types undergo changes resulting in loss of transmission due to the sun's ultraviolet radiation. The effect of dirt and dust on collector glazing appears to be small, and the cleansing effect of an occasional rain seems to be adequate to maintain the transmittance within 2 to 4% of its maximum value.

The absorptivity of the absorber plate also varies with the angle of incidence (Figure 8-3). The common choice for coating the absorber plate is a flat, black oil-based paint, with an appropriate thin undercoat of primer. The primer should be of the self-etching type to avoid the peeling off of the paint after a few years of operation and the paint must be etched on. In any case, collector absorptance should not be less than 90%. As with the previous case concerning absorption of solar radiation in the glass, a reduction in the absorptance of the black absorber does not lead to a corresponding reduction in the system performance. It has been found that a 1% decrease in the absorptance will result in only about a $\frac{1}{2}$% decrease in useful solar output. Again, this is due to lower heat losses in the collector since the system is running somewhat cooler.

The choice of material to use for the absorber plate is a serious one. Copper (Cu), steel, and aluminum (A1) have been by far the most comon choices. No definitive answer can be given, since under proper conditions all three will work satisfactorily. Copper and some types of stainless steel have the advantage that they can be used with untreated tap water, whereas neither aluminum nor carbon steel should be used in this environment under any circumstances. In the case of copper, flow velocities must be chosen to avoid corrosion due to erosion. In Table 8-1 some of the useful characteristics of the three common materials are compared for the case of plate and tube absorber combina-

TABLE 8-1
Comparison of Absorber Plate Materials[a]

ITEM	COPPER	STEEL	ALUMINUM
Density (lb/in.^{3})	0.32	0.28	0.10
Thermal conductivity (Btu/hr-ft-°F)	226	27	122
Plate thickness for equal performance (in.)	0.016	0.064	0.032
Weight per square foot for equal performance in pounds (6-in. tube spacing for Cu and Al; 4-in. spacing for steel)	1	3.9	0.6
Flow correction factor, F' (2-in. tube spacing)	0.996	0.940	0.960
Energy cost per pound (Btu)	47,400	11,600	116,690
Energy cost per square foot (Btu)	47,000	45,200	70,000

[a] Data for this table obtained from articles in the May 1977 issue of *Solar Age* magazine. All data are presented in English units.

tions. In the computation of the energy used for the steel absorber plate, the tube length was assumed to be 50% greater than that for the other two materials because of reduced tube spacing required to give equal performance (reflecting the low thermal conductivity of the steel). Steel sheet could be used in a bonded style as in Figure 8-2a. The resulting

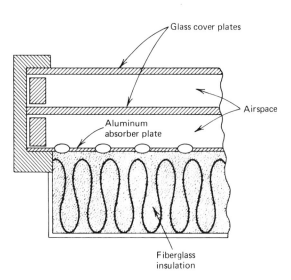

FIGURE 8-4 Schematic illustration of the edge details of a flat-plate collector.

large contact area would reduce greatly the importance of the low thermal conductivity of steel. The largest fraction of the energy cost for the collector comes in the manufacturing of the absorber plate. Regardless of the choice of absorber material, the energy used in the manufacture of the flat-plate collector is approximately the same as the energy that will be collected by the solar system in one year of operation.

The construction details of flat-plate collectors vary considerably; examples can be found by obtaining the flat-plate collector literature from the many commercial companies now in this business. Figure 8-4 illustrates one type of arrangement of glazing, absorber plate, tubes, and insulation for a solar collector. The optimum spacing between glazings has been determined both experimentally and analytically to be about 1.0 to 1.5 cm.

C. FLAT-PLATE COLLECTOR HEAT LOSSES

There are three different physical mechanisms by which heat may be transferred from one place to another. These processes are conduction, convection, and radiation. Conduction is the transfer of heat through a quantity of material due to the molecular motion. For example, if a rod is heated at one end, the molecules there vibrate faster, with the result that energy is transferred down the rod because of successive molecular collisions. Metals are good conductors of heat, whereas materials such as glass, ceramics, wood, and gases are poor conductors of heat. Convection is the process whereby a warm liquid or gas moves from one place to another, thereby effecting a transfer of energy. We have already considered radiation, which involves the transfer of energy by means of electromagnetic waves.

In a flat-plate collector the principal heat-loss mechanisms are the following:

1. Conduction loss from the back of the absorber plate through the insulating material.

2. Conduction losses through the side of the collector.

3. Convection losses upward through the cover plates.

4. The upward radiation loss.

The operation of a flat-plate, solar-thermal collector can be described with the aid of Figure 8-5. Above the absorbing surface, spaced about 1 cm apart, are one or more cover plates (usually glass) whose function is to reduce upward heat losses. Most of the sun's rays ($\cong 85\%$ for one cover plate) are transmitted through and are absorbed by the black surface. Typically, 92 to 96% of the radiation actually incident on the

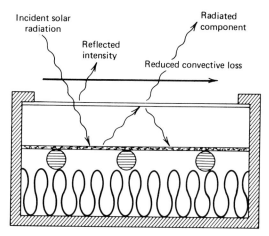

Incident solar radiation

Radiated component

Reflected intensity

Reduced convective loss

FIGURE 8-5 Schematic illustration of the green-house effect. The major feature is the reduced convective loss at the top cover plate.

blackened plate is absorbed. As the metal plate warms up it emits radiation predominantly in the far-infrared wavelength region. The peak wavelength in this region was calculated in Chapter 6 using the Wien displacement law, which says that the peak of a blackbody radiation spectrum is related to the temperature by $\lambda_m T = K$. Although glass is transparent to visible light it is essentially opaque to the wavelengths larger than 3000 nanometers (nm). Thus, in a sense the sun's energy is trapped in the enclosure by the combination of a black absorber and the glazing that covers it. The cover plates warm up to a temperature between that of the hot absorber plate and the outside environment. This results in a substantial decrease in the radiation emitted by the collector back to the outside world. It will be shown later that while the layers of glass (or plastic) absorb and reflect some of the incoming solar intensity, the reduction in heat loss outweighs the loss in transmission, except at very low operating temperatures. Some of this heat loss reduction is due to the interception of infrared radiation by the glass, but a significant reduction also comes from the formation of blankets of relatively stagnant air. With proper spacing of the glazing, natural convection effects are greatly reduced. For example, some plastics used as covers for solar collectors are relatively transparent to the infrared radiation, but still insulate effectively because of the reduction in convective losses. The topmost cover plate greatly reduces heat losses due to direct wind convection, which is by far the dominant heat loss mechanism when compared to the situation where no cover plate is used. A further reduction in the heat loss can be obtained through the use of a selective surface that has a low emissivity for the infrared and high absorptivity for the incident solar radiation spec-

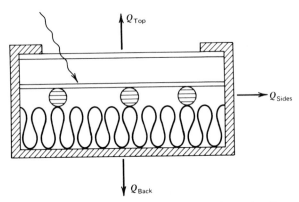

FIGURE 8-6 Schematic drawing of the heat losses in a flat-plate solar collector. The total heat loss is the sum of the losses through the sides, back, and top of the collector. In a properly designed collector the loss through the top is dominant.

trum. The division of the heat losses from a flat-plate collector is shown schematically in Figure 8-6. The rate of energy loss through the back insulation per unit area is directly proportional to the thermal conductivity K and the difference of temperature between the absorber plate and the environment in back of the collector frame, and is inversely proportional to the thickness, D, of the insulation. This is mathematically expressed as

$$Q_{Back} = \frac{K}{D}(T - T_{Back}) = \frac{K}{D} \Delta T_{Back} \qquad (8\text{-}1)$$

where T is the temperature of the absorber plate and T_{Back} is the temperature at the back of the collector. Since it is easy and economical to provide good insulation on the back of the solar-thermal collector, this heat loss is usually small compared to the losses by convection and radiation of the top surface. Typically, 7 to 10 cm of fiberglass insulation is sufficient for this purpose. Equation 8-1 can be written in the following form:

$$Q_{Back} = U_{Back} \Delta T_{Back} \qquad (8\text{-}2)$$

where U_{Back} is called a heat transfer coefficient. The heat transfer coefficient, U_{Back}, is then a measure of the heat loss through the back insulation. A well-designed collector will have U_{Back} as small as possible, consistent with the economics of the situation. For most collector designs, the heat losses through the sides and the back are small compared to the losses through the top cover plate. A useful approximation is that the heat losses through the sides and the back are about 10% of the upward heat loss.

The upward heat loss is schematically illustrated in Figure 8-7,

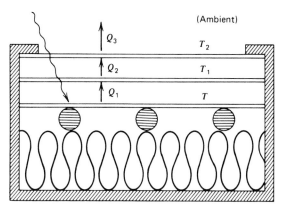

FIGURE 8-7   Cutaway through a flat-plate solar collector indicating the transfer of heat from the hot absorber plate. At equilibrium each of the heat flows—Q_1, Q_2, and Q_3—are identical.

showing the heat losses Q_1, Q_2, and Q_3 between the absorber and second cover plate, between the second and first cover plate, and between the first cover plate and the outside world, respectively. (This simple picture neglects minor corrections such as the energy absorbed from the incident solar radiation by the cover plates.) The upward heat loss is mainly due to convection and radiation. Between the black absorber plate and the adjacent cover plate, and between the two cover plates, the convection and radiation losses are roughly the same. To reduce the interior convection, some flat-plate collectors have been constructed with a honeycomb-type of cellular structure. In cylindrical evacuated tube glass collectors, this convection loss is eliminated. For the cover plate that is in contact with the outside world the dominant loss mechanism is convection. At a wind speed of 9 to 15 km/hr this convection loss can exceed the radiative one by more than a factor of 5. Consequently, flat-plate collectors are usually constructed with at least one glazing.

It was shown in Chapter 6 that the total energy radiated by a hot object is proportional to the fourth power of the absolute temperature. Two surfaces at different temperatures will then have a net amount of energy transferred between them. The calculation of this energy transfer between two parallel plates is quite straightforward but, since it involves summing up infinite series, is beyond the level of this text. The important result is that the net radiative transfer between the hot absorber plate and a cover plate is given approximately as

$$Q_{Rad} = \varepsilon_p \sigma (T^4 - T_1^4) \tag{8-3}$$

where ε_p is the infrared emissivity of the absorber plate and T, T_1 are the absolute temperatures of the absorber and cover plate, respectively. The emissivity of the cover plate is assumed to be unity for the long-wavelength radiation emitted by the hot absorber surface. If T and T_1 differ only slightly, which is the case for flat-plate collectors, then $T^4 - T_1^4$ is approximately proportional to $\Delta T = T - T_1$. In this case,

$$Q_{Rad} = U_{Rad}\Delta T \qquad (8\text{-}4)$$

which is a form just like that found in Eq. 8-2. The convective loss can be approximated by a similar form so that the total amount of heat transferred from one surface to another can be written as

$$Q = U\Delta T \qquad (8\text{-}5)$$

for each pair of surfaces (considering the outside world as a surface). Finally we note that if we know the individual U's, we can directly calculate an overall heat transfer coefficient U_{Top}:

$$Q = U_{Top}\Delta T_{Top} \qquad (8\text{-}6)$$

where $\Delta T_{Top} = T - T_0$ is the difference in temperature between the absorber plate and the outside world. The calculation of U_{Top} is one of the important tasks facing an engineer interested in designing a flat-plate collector.

The operation of a flat-plate collector can be simply described using Figure 8-7. Assume that the collector has come to equilibrium so that all temperatures and heat flows have assumed constant values. The incoming solar radiation is absorbed by the black plate, which heats up, reaching an equilibrium temperature T (assumed uniform in our treatment), which is determined by the rate at which heat is carried away by the water and by the rate at which heat is radiated and convected to the adjacent cover plate. (Remember that the radiation emitted by the absorber plate is of very long wavelength and is almost entirely absorbed by a glass cover plate or partially absorbed by a plastic cover plate.) The cover plate comes to an equilibrium temperature T_1 lower than T. Heat is then transferred from this cover plate to the next, which comes to an equilibrium temperature T_2 lower than T_1, but is still somewhat higher than the outside ambient temperature T_α. This top cover plate then loses heat to the outside world by radiation and convection, with the latter term strongly dominant when there is a wind.

It is now apparent why it is advantageous to utilize cover plates. The heat loss from the collector is reduced through the use of cover plates because the temperature of the surface in contact with the outside atmosphere is reduced. Besides cost, the major negative feature of cover plates is that the transmission through them is not perfect, so that a compromise must be struck between the reflection and absorption losses, which increase with the number of plates used, and the upward heat loss, which decreases as the number of cover plates increases.

The effect of increasing the number of cover plates is graphically

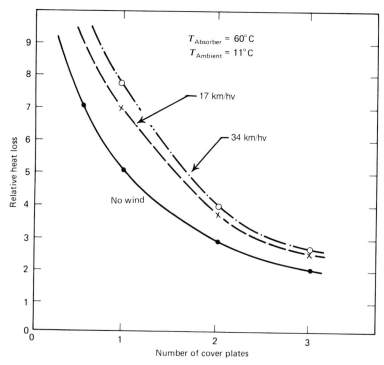

FIGURE 8-8 Showing the reduction in upward heat loss as the number of cover plates is increased.

portrayed in Figure 8-8, which shows the relative heat loss of a flat-plate collector as a function of the number of cover plates. With an assumed temperature of 60°C for the absorber plate and 11°C for the outside ambient temperature, it is seen that the convection losses due to wind striking the outer cover plate are still quite large for one cover plate. When two cover plates are used, the heat loss is strikingly reduced, and the sensitivity to wind is also greatly reduced. Only a minor improvement is obtained when three cover plates are introduced; in practice the number of cover plates is almost never greater than two.

D. COLLECTOR PERFORMANCE

The useful heat output from a flat-plate solar-thermal collector is equal to the difference between the heat absorbed and the heat loss by the collector and can be written quantitatively as

$$Q_u = \text{heat absorbed} - \text{heat lost}$$
$$= \tau\alpha\Phi_0 - U_L\Delta T \qquad (8\text{-}7)$$

where U_L is the sum of the heat transfer coefficients through the top, sides, and back. This result simply expresses conservation of energy as it relates to the flat-plate collector. The incident solar radiation is Φ_0, and ΔT is the temperature difference between the absorber plate and the outside ambient temperature. (It is important to always remember that ΔT is the difference between the average absorber plate temperature and that for the outside world and is not to be confused with the temperature rise of the heat-exchange medium as it crosses the hot absorber plate.) In the first term of Eq. 8-7, $\tau\alpha$ is the product of the resultant transmittance through the collector cover plates with the absorptance by the black absorber plate averaged over the angles of incidence of the incoming solar radiation over some period of time. With no cover plates τ is one, and for low ΔT the useful heat collected will be greater than with cover plates. However, U_L will be much larger in this case, particularly if there is any wind, and as ΔT increases the useful heat collected will be much larger if cover plates are used.

It is instructive to rearrange Eq. 8-7 by dividing through by the incident solar flux Φ_0. The ratio of Q_u to Φ_0 gives the fraction of the incident insolation that is captured by the solar collector and is termed the instantaneous collector efficiency η.

$$\eta = \frac{Q_u}{\Phi_0} = \tau\alpha - \frac{U_L\Delta T}{\Phi_0} \tag{8-8}$$

Typically, about half of the incident energy is absorbed and converted into useful energy; this means that the efficiency η is 0.50, or 50%. When $U_L\Delta T$ is greater than $\tau\alpha\Phi_0$, the incoming solar radiation can no longer provide for the collector losses. This will occur when the collector is warm, the outdoor air is cold, and the incoming radiation intensity is low. Thus, when $\eta = 0$, the heat losses are equal to the energy absorbed and no useful energy is obtained.

Note that whenever $U_L\Delta T$ is small compared to Φ_0, η is roughtly $\tau\alpha$, the maximum efficiency attainable. Clearly, operating the collector so that the absorber plate temperature is close to the ambient gives the highest efficiency. Inasmuch as the efficiency of a solar-thermal collector is strongly affected by the difference between the temperatures of the absorber plate and the outside atmosphere, it is desirable to hold the absorber plate temperature low to obtain high efficiency. But at the same time a reasonably high temperature is needed. For this reason the inlet to the collector is usually connected to the lowest temperature point of the storage medium.

Figure 8-9 illustrates how the performance of a flat-plate collector goes down as the collector temperature increases and as the insolation Φ_0 decreases. This demonstrates again the advantage of using two cover plates when ΔT is large. The efficiency drops off most rapidly with ΔT

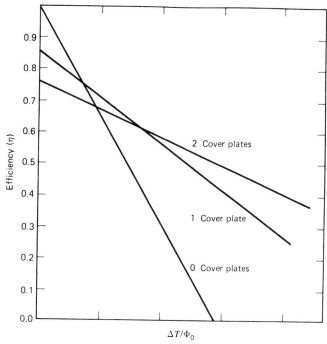

FIGURE 8-9 Plot showing how the efficiency of a solar collector varies as a function of the temperature difference between the absorber plate and the outside atmosphere as well as the intensity Φ_0 of the incident solar radiation. The heat transfer coefficient U can be obtained by calculating the slope of the efficiency curve.

for the case of no cover plates. All of the incident solar energy reaches the absorber plate when no cover plates are used because of no transmission loss. However, at higher values of ΔT, the convective heat loss dominates and the addition of cover plates greatly reduces this loss. From Eq. 8-7 we can also see how to empirically obtain U_L. By measuring Q_u, ΔT, and Φ_0 under varying conditions one can trace out the collector efficiency curve. The slope of the straight line through these points then gives U_L, without recourse to any additional calculational procedures.

The use of efficiency curves to compare different collectors must be viewed with caution. Some collectors have good efficiency curves when evaluated at noon, but fall off rapidly as the rays from the sun make larger angles with the normal to the collector. In fact, one is usually interested in the average performance over an entire day. Similarly, the efficiency measured during the winter months is most relevant for space and water heating applications, whereas the performance in summer is most relevant for cooling purposes.

There are many ways of improving collector performance:

1. Coat the black absorber plate with a special surface called a selective surface. This can significantly reduce the upward radiative loss in the collector. (Selective surfaces are discussed in more detail in the next section.) However, the cost of such surfaces is high, and their lifetime expectancy is uncertain.

2. Apply an antireflection coating to the surface of the collector glazing. The reflection loss (about 16% for a double-glazed reflector) can be reduced by a factor of 2 to 5. Again, these coatings are costly and not durable.

3. Use evacuated glass collectors to reduce the conduction and convection losses. To do this requires special efforts to withstand atmospheric pressure. Presently glass collectors are quite expensive and their use appears limited to high-temperature applications such as solar cooling and process heat.

4. Enhance the amount of radiation entering the collector by using reflectors. This approach is discussed later.

We conclude this discussion of collector performance by noting that thus far the temperature of the absorber plate was assumed to be constant over its entire surface. This is far from true. In the type of collector design illustrated in Figure 8-7, the temperature of the absorber plate is hotter between the tubes carrying water than it is at the tubes themselves. This nonuniform distribution is essential so that there will be a flow of heat from the hot absorber plate into the cooler heat transfer fluid. It is this heat that is carried away by the fluid that is used to provide practical space heating and cooling. The temperature distributions between pipes for this style of flat-plate collector design are schematically illustrated in Figure 8-10. The temperature also varies along the flow direction as the fluid warms up as it traverses the plate. This results in a roughly linear distribution of temperature from inlet to outlet. Since the average temperature across the absorber plate for such a linear distribution is just

$$T_{av} = \frac{T_{in} + T_{out}}{2} \tag{8-9}$$

this definition is often used in calculating $\Delta T = T_{av} - T_a$, with T_{av} computed from Eq. 8-9. As the inlet and outlet water temperatures are commonly monitored, this usage has the practical advantage of defining the collector efficiency in terms of easily measured quantities. Alternatively, $\Delta T = T_{in} - T_a$ is often used in Eq. 8-7.

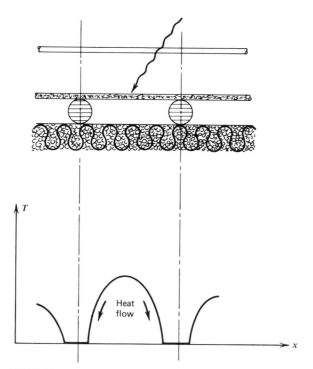

FIGURE 8-10   Cutaway drawing of a solar thermal collector. This illustrates roughly the temperature distribution between the fluid-carrying pipes. The temperature of the absorber plate is highest midway between the water tubes.

This nonuniform distribution of temperature across the collector causes a modification to Eq. 8-7. An extensive mathematical analysis shows that Eq. 8-7, for the net heat collected, becomes

$$Q_u = F_R(\tau\alpha\Phi_0 - U_L\Delta T) \qquad (8\text{-}10)$$

where $\Delta T = T_{in} - T_a$ and F_R is the ratio of the actual heat collected by the solar collector to the heat collected if the entire solar collector were at the inlet fluid temperature. For most reasonable collector designs of the liquid type, this heat removal factor F_R is of the order of 0.85 to 0.90. Collector designs using air as the heat-exchange fluid often have F_R factors that are considerably smaller. However, the improved heat transfer from storage to the building being heated and the lower input temperature to the collector, obtained with air systems, overcome the disadvantage of having a lower efficiency for the useful heat collected by the flat-plate collector.

E.   SELECTIVE SURFACES

It has been noted before that one way to reduce the upward heat loss in a solar collector is to use an absorber surface that is black to the incident solar radiation, but emits very weakly in the long-wavelength region. Figure 8-11 shows the radiated intensity as a function of wavelength for both the solar spectrum and for a typical absorber plate (about 93°C or less). It is seen that there is essentially no overlap between the two distributions.

A selective surface is one that absorbs the incident solar energy, but suppresses its thermal radiation. Thus, if α is about 0.94 in the visible and near-infrared regions (300 to 1200 nm) and is less than or equal to 0.1 for wavelengths greater than 1200 nm, a large reduction in the radiative energy loss will be obtained. This is illustrated by looking at the behavior of a nearly ideal selective surface as shown by the dashed line in Figure 8-11. The absorptance is almost unity below 1200 nm and almost zero above this wavelength. Since $\alpha + r = 1$, another way of looking at this is to say that the reflectance is almost zero below 1200 nm and almost one for wavelengths greater than this.

The principle of operation of a selective surface is illustrated in Figure 8-12. The copper absorber plate is covered by thin layers of zinc and silver in this type of selective surface. The solar radiation incident on the surface is absorbed by the layer of zinc. Thus the absorptivity in the visible and near-infrared wavelength regions is almost unity. However, as shown in Figure 8-12b, radiation in the infrared region (corresponding to emission from a hot absorber plate) passes through the thin layer of zinc and is reflected by the thin layer of silver. Thus, the silver layer acts as a mirror for long-wavelength radiation. As

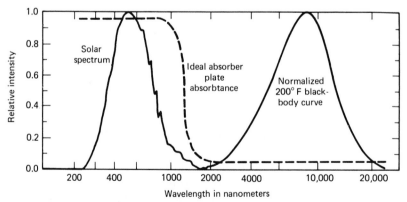

FIGURE 8-11   A diagram of the relative intensity distributions for the solar spectrum and for a typical absorber plate in a flat-plate collector. An excellent selective surface would be almost black in the solar region and have almost no absorptance in the long-wavelength region.

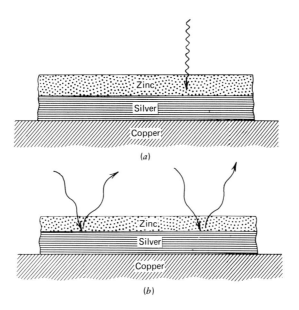

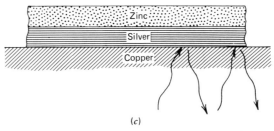

FIGURE 8-12 Illustrating the behavior of a selective surface. See text for a discussion of the principles of operation.

shown in Figure 8-12c, the infrared radiation from the hot copper absorber is mostly reflected back into the absorber, and thus cannot escape. The net result of all this is a surface with low emissivity in the infrared wavelength region. The materials for the selective surface illustrated in Figure 8-12 were chosen only to illustrate the principle involved. A common practical selective surface uses black chrome on bright nickel electroplated over a copper substrate.

Detailed simulation calculations of actual solar systems have shown that significant improvement can result through the use of selective surfaces. For example, a single-glazed solar system in Bismarck, North Dakota with a selective surface (emittance of 0.10 for the far infrared) would require only 62% of the collector area needed for a single-glazed collector with no selective surface. These calculations assume that the solar system provides 75% of the total space heating load. For Medford, Oregon, similar calculations have shown that the selective surface system requires only 70% of the collector area needed without the selective surface. The corresponding percentage for sunnier areas is higher with a value of 84% for Phoenix, Arizona. In all cases, performance was actually reduced by the use of double glazing over a selective surface as compared to single glazing over a selective surface. This is because the increased absorption and reflection caused by the second layer of glass results in a greater reduction in system output than the corresponding gain resulting from decreased convective and radiative losses. In summary, if the economic costs are justifiable, it is advantageous to use a selective surface with one glass cover plate. If two cover plates are used then no selective surface is the best choice.

F. CONCENTRATING COLLECTOR SYSTEMS

While space and water heating systems most often accept and utilize the incoming solar radiation as is, the development of efficient devices to produce mechanical power or electricity requires concentrating optical systems. The subsequent increase in technical complexity is more than offset by the higher temperatures available in the absorbing systems and the greatly reduced thermal losses. Concentrating systems can be divided into types defined in terms of the number of degrees of freedom of the tracking mechanism:

1. Nontracking devices are characterized by low-concentration ratios (the ratio of the flux density of the absorber to that of the incident solar radiation) in the range of 1 to 10. Examples of this approach are the use of reflecting mirrors with flat-plate collectors (Figure 8-16) and the so-called "horseshoe-crab collector" (Figure 8-17), which concentrates the incoming solar radiation by a factor of 10 or so without following the sun. In an optical sense this approach is known as a nonimaging concentrator.

2. Collectors with cylindrical symmetry, such as the parabolic trough concentrator (Figure 8-18), need only be tracked with mechanisms having one axis of rotation. In this case the incident radiation is usually concentrated by a factor of 10 to 50.

3. Finally there are systems with quite high concentration ratios that require tracking mechanisms having 2 degrees of freedom. Naturally, the tracking precision needed is much greater for these systems than for single-axis tracking systems. These are compound curva-

ture systems, such as the parabolic reflecting mirrors used in a car's headlights. An important example of this concept is the solar tower. In this approach, a heat-exchange fluid located on top of a tall tower is heated by the solar radiation reflected from thousands of individual mirrors.

Concentrating devices increase the solar flux density over that normally obtained when no concentration is utilized. This concentration is achieved through the proper orientation of refracting or reflecting surfaces. There are a number of good reasons for wanting a high solar flux density as the receiver. The efficiency of all devices operating on a thermodynamic cycle improves as the temperature of the heat source rises. High-quality energy in the form of electricity is needed and its production by thermal processes requires the use of an engine operating through some type of thermodynamic cycle. To achieve thermodynamic efficiencies greater than 40%, operating temperatures above 1000 K will probably be required.

Higher flux densities at the receiver (and the subsequent higher operating temperatures) mean that the thermal losses per unit area will be increased. This increase in a well-designed system is more than compensated for by the greatly reduced receiver area required (for a given total amount of energy collected) when compared with a nonconcentrating system. The overall result is to effect a very large reduction in the thermal heat losses with concentrating collector systems.

The cost per kilowatt hour should also be decreased. Cost reductions will occur because the reflecting surfaces are much less expensive than the more complex receivers. Counteracting this is the increased costs imposed by the more stringent optical requirements of focusing devices. The generation of electricity also requires an expensive receiver, turbine system.

The utilization of concentrating systems introduced losses not present with nonconcentrating systems. Most of the diffuse component is not collected in focusing devices, as only the radiation coming from the region of the sky quite near the sun is collected. Imperfections in the reflecting (or refracting) surface cause additional losses. Also, in many approaches, such as solar towers, the need to avoid shadowing has the result that only a fraction of the total area encompassed by the reflectors is actually effective.

For precise tracking, some type of feedback control device is employed. Commonly, the reflecting system (occasionally the receiver) is moved at a standard rate by a clock-driven mechanism to keep the sun, even when obscured by clouds, within the desired range of acceptance of the system. Precision adjustments are then accomplished by using a feedback control actuated by some type of sun-sensing device.

G.   REFLECTING OPTICS

The preponderance of focusing devices utilize spherical or parabolic reflecting mirrors in some type of geometry. The elementary principles of these devices can be illustrated by considering the spherical mirrors shown in Figure 8-13. In addition to having no transmission losses, the image from a spherical mirror is superior in some respects to that from a lens, notably in the absence of chromatic effects due to the different bending that occurs whenever light rays are refracted (inasmuch as light is made up of a spectrum of wavelengths).

A quantitative description of the optics of spherical mirrors is easily described using Figure 8-14. This figure illustrates how the image of a particular object can be obtained using a simple graphical procedure. Point F is the focal point of the mirror and C is the center of curvature. Any line such as 1 which is drawn parallel to the optical axis must reflect back through F as shown. Any ray drawn from the object through F must be reflected back parallel to the optical axis. Finally, a ray such as 3 drawn through the center of curvature will be reflected back along its original path. In this way, the image is located a distance S from the lens.

We can also obtain the image distance S with the aid of the following equation:

$$\frac{1}{O} + \frac{1}{S} = \frac{1}{F} = \frac{2}{R} \qquad (8\text{-}11)$$

where O is the object distance, F is the focal distance, and R is the radius of curvature. It is customary to adopt a convention in which R is

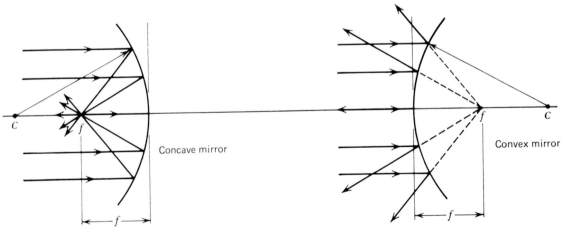

FIGURE 8-13   Illustration of the focal points for concave and convex spherical mirrors. The point C marks the center of curvature of the mirror of radius of curvature, R. For spherical mirrors R = 2f.

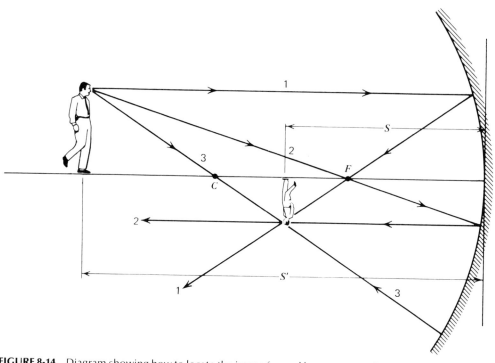

FIGURE 8-14   Diagram showing how to locate the image formed by a concave mirror using the method of principal rays.

positive for a center of curvature located to the left of the mirror in Figure 8-14.

EXAMPLE 1

An object 4 cm high is located 20 cm in front of a concave mirror with a radius of curvature of 30 cm. We find S with the aid of Eq. 8-11.

$$\frac{1}{20} + \frac{1}{S} = \frac{2}{30}$$

or

$$S = 60 \text{ cm}$$

By graphical construction, interested students can easily convince themselves that the image is inverted and is larger than the object.

In practice spherical mirrors are satisfactory only if just a small segment of the actual spherical surface is used as a reflector. If too large a portion of the spherical surface is utilized, then parallel rays from a considerable distance off the principal axis come to a focus at points quite close to the axis. This focusing defect, called spherical aberration, is shown in Figure 8-15.

Numerous methods of reducing spherical aberrations have been devised. When a *corrector* plate is located at the center of curvature, the popular *Schmidt* telescope system is obtained. Another approach is to shape the mirror in the form of a parabola. A parallel beam of light incident on a parabolic reflector, no matter how large the aperture, is all focused at one point. This is the type of mirror used in astronomical telescopes, search lights, and automobile headlights, as well as in many solar energy applications.

A simple example of the use of reflecting optics with flat-plate collectors is the use of planar reflectors to enhance the amount of solar radiation captured. We illustrate this for winter heating operation. In areas where there is low insolation during the winter, a standard solar collector may only be turned on to collect useful energy for a few days each month. This is the case for cloudy areas of the country such as the Pacific Northwest. Inexpensive reflectors can be used to enhance the

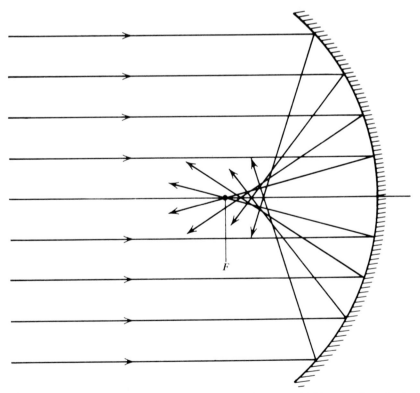

FIGURE 8-15 Diagram illustrating spherical aberration. Rays incident on the spherical mirror at a considerable distance from the axis are focused at points in front of the point *F*.

light collection of flat-plate collectors, thereby greatly improving their overall performance.

The concept of using reflectors with flat-plate collectors is not new, having been used as early as 1911 by F. Shuman in conjunction with a tiltable collector array for a solar power plant at Tacomy, Pennsylvania. A modest reflector was used in one of the early M.I.T. houses and also in the solar house designs of H. E. Thomason. The first person to make use of a large reflector in an optimally oriented reflector-collector system was H. Mathew of Coos Bay, Oregon. The striking feature of this house is the nearly vertical inclination of the collector rather than the customary tilt angle of 55 to 60° for a collector located at 45°N latitude. A description of this house is given in Chapter 9.

A qualitative understanding of why a nearly vertical collector is best when a reflector is utilized can be obtained with the help of the diagrams in Figure 8-16. We assume that the sun's rays are incident at some angle θ (from the vertical). This angle varies with the time of day

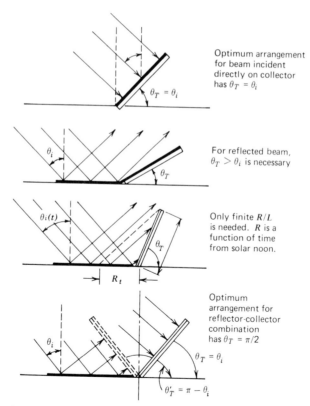

Optimum arrangement for beam incident directly on collector has $\theta_T = \theta_i$

For reflected beam, $\theta_T > \theta_i$ is necessary

Only finite R/L is needed. R is a function of time from solar noon.

Optimum arrangement for reflector-collector combination has $\theta_T = \pi/2$

FIGURE 8-16 Illustration of the reflector-collector geometry showing why a near vertical collector is the optimum arrangement.

and year, but detailed calculations show that this qualitative picture is valid. The top diagram shows that the optimum geometry for a simple collector has the incident beam radiation striking the collector normally. In this case the collector tilt angle θ_T equals the angle of incidence θ_i. The second diagram shows that for the reflected beam to strike the collector, the collector tilt angle must be greater than θ_i (for a horizontal reflector). The third diagram illustrates the fact that only a finite reflector length is needed for any given angle of elevation of the sun above the horizontal plane. The bottom drawing shows that the collector orientation for optimum gathering of the reflected intensity only occurs when $\theta_T = \pi - \theta_i$. If the reflected intensity and that directly incident on the collector are roughly equal, then the best collector orientation is the average of the optimum angles for the two individual contributions, leading to the conclusion that the collector should be oriented vertically.

Detailed analysis shows that the optimum reflector orientation for winter heating at 45°N latitude is about 5 to 10° downward, with an angle of 100° between the reflector and collector planes. In this geometry, the amount of light gathered by the reflector system in the winter is about 55 to 60% greater than that collected by a standard flat-plate collector oriented at 60°.

How large should the collector be? For flexibility in architectural design, the shortest possible reflector length is desirable. Detailed calculations have been made for the case where the reflector was oriented 5° downward and the collector was oriented 85° upward from the horizontal plane. This analysis shows that the amount of light gathered by a system with a reflector length R, equal to twice the collector height L, is only about 4% less than that obtained with an infinite reflector. These same calculations show that losses, due to having a reflector with a finite width, can be kept below 3% (comparing again with an infinite reflector) if the reflector width to collector height ratio, W/L, is larger than 3. (The reflector width is always assumed to be the same as the width of the collector.)

H. EXAMPLES OF CONCENTRATING SYSTEMS

The most commonly used concentrating systems involve the utilization of reflecting optics, in much the same manner as most astronomical telescopes. Before going on to describe a few of these systems, it will be useful to first define several widely used terms. The aperature area of the collecting system is the total area intercepting solar radiation. The area that receives the concentrated radiation (after reflection) is known as the absorber or receiver area. Finally, the concentration ratio (CR) is defined as the ratio of the net collecting aperature area to the net receiver area.

There are some general limits on the amount of concentration that can be achieved. For a two-dimensional geometry (cylindrical or trough-like symmetry), application of the second law of thermodynamics to the problem of radiative heat transfer leads to the result that

$$CR_{cyl}{}^{max} = \frac{1}{\sin \theta_m} \qquad (7\text{-}12)$$

In this equation $CR_{cyl}{}^{max}$ is the maximum concentration ratio achievable by devices with cylindrical symmetry and θ_m denotes the coverage of one-half of the angular zone within which radiation is accepted by the receiver. The actual angle of acceptance is then $2\theta_m$. This angle is illustrated in Figure 8-17 for the Compound Parabolic Concentrator system which is described later in this section.

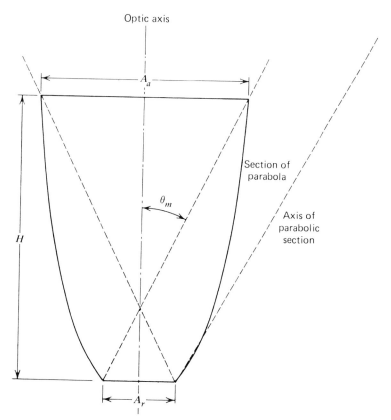

FIGURE 8-17   Schematic drawing of a CPC solar collection system. The solar radiation is incident on the top of the system and collected along the surface, A_r, at the bottom.

The corresponding result for dish-type concentrators, which have symmetry about a line, is found to be

$$CR_{Sph}{}^{Max} = \frac{1}{\sin^2 \theta_m},$$ (7-13)

where $CR_{sph}{}^{Max}$ is the maximum concentration ratio achievable by devices with symmetry about a line, such as the solar tower or a dish-type concentrator. The maximum theoretical limit on concentration is then set by the size of the solar disk which subtends 32 min of arc to an observer on earth. This gives θ_m = 16 min and leads to $CR_{cyl}{}^{Max} \approx 240$ and $CR_{Sph}{}^{Max} \approx 57,000$. In actual working systems the concentrations achieved will be much less because of design constraints, tracking errors, and imperfections in the reflecting (or refracting) surface.

Not much diffuse radiation will be accepted by a concentrating system. Assuming an isotropic distribution or radiation, the same type of general thermodynamic reasoning as above leads to the conclusion that a fraction 1/CR of the diffuse component is collected. Because the diffuse component is not isotropic, but instead is somewhat concentrated in the region near the sun, an even larger fraction of the diffuse component will be collected. In some cases this can result in a significant improvement of the solar energy collected over that obtained from just the direct component.

Conventional concentrating systems involve some degree of tracking of the sun. However, a fixed-collector system can provide some seasonal enhancement of the energy collected by flat-plate collectors. For example, as we saw in the last section, the use of simple fixed planar mirror reflectors located in front of an almost vertical collector, when properly oriented, can provide up to a 50% boost in the collected solar energy, averaged across an entire winter heating season, over that obtained with the usual flat-plate collector oriented at the standard rule-of-thumb tilt angle of latitude plus 10 to 15°. Planar mirrors located on the sides of flat-plate collectors oriented almost horizontally have been suggested, but these require daily adjustment of the mirrors without any significant improvement in the concentrations achieved with the fixed mirror, tilted collector approach.

An example of a moderately concentrating system that may only require adjustment one to two times per month is the Compound Parabolic Concentrator (CPC). This approach was introduced by Winston in 1974, originally as a new design concept to collect Cherenkov radiation used for particle identification and timing purposes in nuclear physics experiments. The principle of operation is illustrated in Figure 8-17. The basic idea is that any ray of light entering the collector that makes an angle less than or equal to θ_m with the optical axis will strike the receiver area. This is a nonimaging type of concentrating system, that is, the incident light is not focused to a point. For moderately concentrating systems that is not a practical disadvantage, and this approach has the advantage of being character-ized by the highest theoretical concentration allowed, CR = $1/\sin \theta_m$.

The CPC is an example of a concentrating system that requires very little tracking of the sun. More commonly, reflecting systems for concentrating devices require daily tracking of the sun across the sky. The simplest tracking mechanism is known as a single-axis tracking system as it involves a straightforward rotation about an axis. Single-curvature concentrating devices generally are constructed with cylin-drical symmetry and are tracked by mechanisms having 1 degree of freedom. The moderate concentration ratio obtained, usually in the 10 to 50 range is suitable for heating fluids to temperatures in the range of 100 to 500°C. The basic ideas involved here are certainly not new, having been utilized by a number of others in the early part of the 20th century, such as the Shuman and Boys power plant in Egypt that developed 50 hp for an irrigation project. These early engines all suffered from very low overall efficiencies that were only of the order of a few percent.

Modern technology offers the possibility of a number of improve-ments over earlier solar designs. Moderate concentration ratios can be obtained through the use of well-constructed parabolic reflectors. These can either track the sun and focus the sun's rays on a stationary receiver or, alternatively, a stationary reflector can be combined with a movable receiver. Losses in the receiver can be minimized through the use of techniques such as an evacuated inner element combined with a selective surface. This eliminates convective losses and greatly reduces radiative emission in the infrared from the hot fluid. An example of one of these systems is shown in Figure 8-18. Further discussion of single-axis (and double-axis) tracking systems is contained in Chapter 10.

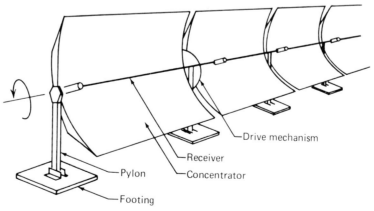

FIGURE 8-18   The parabolic trough concentrator is an example of a single-axis tracking system. During the day the parabolic reflector is rotated about its axis of symmetry as indicated.

REFERENCES

1. H. C. Hottel and D. D. Erway, "*Collection of Solar Energy*," Chapter 5, in *Introduction to the Utilization of Solar Energy*, A. A. Zarem and D. D. Erway, eds. (New York: McGraw-Hill, 1963).

    For those interested in a comprehensive treatment of flat-plate collectors, this concise article is an excellent place to start. The effect on the collector efficiency of the materials used, the solar energy input, the number of glazed surfaces, and other parameters are discussed and results presented for typical cases.

2. J. A. Duffie and W. A. Beckman, "Flat-Plate Collectors," Chapter 7, in *Solar Engineering of Thermal Processes* (New York: Wiley, 1980).

    A detailed description of flat-plate collector design. This book is the "Bible" for many different solar energy applications. Should be on the must reading list for anyone interested in the technical aspects of solar energy.

3. F. Kreith and J. F. Kreider, *Principles of Solar Engineering* (New York: McGraw Hill, 1978).

    An excellent engineering text that contains a wealth of technical information. The interested reader needs to have a sound mathematical background.

4. G. Daniels, *Solar Homes and Sun Heating* (New York: Harper & Row, 1976).

    Chapter 6 contains an elementary description of flat-plate collectors along with practical construction details.

QUESTIONS

The figure below is drawn to represent a cross-sectional view of a typical solar collector. Unless otherwise noted the cover plates are made of ordinary glass.

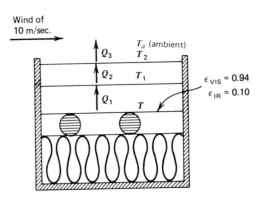

The next 25 questions refer to this figure.
1. Which of the following heat-exchange media is usually not used with this type of solar collector?

(a) Air

(c) Ethylene glycol and water

(b) Water

(d) Oil

2. The most serious problem with using a thin plastic sheet as the glazing material is:

(a) It is structurally too weak.

(b) It absorbs too little of the incoming radiation.

(c) It becomes partly opaque after prolonged exposure to ultraviolet radiation.

(d) It is too expensive for routine use in collectors.

3. The upward heat loss through the top of the collector is typically what percentage of the total collector heat loss?

(a) 5%

(c) 50%

(b) 25%

(d) 90%

4. The percent transmission of the incoming solar radiation through an ordinary glass top cover plate is typically:

(a) 98%

(c) 86%

(b) 94%

(d) 50%

5. The absorptivity of the black absorber plate when covered with a common flat-back black paint is commonly:

(a) 80%

(c) 94%

(b) 85%

(d) 98%

6. The radiative heat loss between the top cover plate and the outside world is proportional to:

(a) $T_1^4 - T_2^4$

(c) $T_2^4 - T_a^4$

(b) $T_1^4 - T_a^4$

(d) $(T_2^4 - T_1^4)^2$

7. Compare the heat loss between the absorber plate and the cover plate with that between the cover plate and the outside world.

(a) The two heat losses are equal.

(b) The heat loss between the cover plate and the outside world is greater.

(c) The heat loss between the absorber plate and the cover plate is greater.

(d) Insufficient information available to decide.

8. Suppose that a 15-km/hr wind is blowing. The dominant heat loss mechanism (or mechanisms) between the absorber plate and the adjacent cover plate is:

(a) Convection.

(c) Radiation.

(b) Conduction.

(d) Convection and radiation.

9. The efficiency of this solar collector is largest at which of the following temperatures for the absorber plate? Assume $T_a = 5°C$.

(a) 20°C   (c) 65°C
(b) 50°C   (d) 95°C

10. When solar radiation strikes the boundary between the air and the top of the cover plate, what percentage of the incident intensity is reflected?

(a) 0%   (c) 8%
(b) 4%   (d) 12%

11. The distance between glazings is typically:

(a) 0.10 cm   (c) 1.0 cm
(b) 0.35 cm   (d) 10 cm

12. Suppose that the absorber plate is coated with a selective surface. Which of the following far-infrared emissivities would offer the greatest reduction in collector heat losses?

(a) 0.05   (c) 0.50
(b) 0.10   (d) 0.95

13. Suppose that the reflectivity of the black absorber surface is 0.06 for visible light. What is the absorptivity of this same surface in the visible region?

(a) 0.06   (c) 0.94
(b) 0.12   (d) 1/0.06

14. For routine operation of the solar collector near 100°C, the optimum number of cover plates would be (assume no selective surface):

(a) 0   (c) 2
(b) 1   (d) 5

15. Suppose that the collector-absorber plate is made of roll-bond aluminum. Which of the following should be used as the heat-exchange fluid?

(a) Water
(b) Air
(c) A mixture of ethylene glycol and water with a corrosion inhibitor.
(d) A mixture of water and finely ground particulate matter

16. Assume that the black absorber plate has been painted with a typical type of flat-black paint. About what percentage of the incident solar radiation is reflected back by the absorber plate?

(a) 94%   (c) 6%
(b) 50%   (d) 1%

17. Assume that $T_1 - T_2$ is about 20°C. How does the magnitude of Q_2 compare with that of Q_1?

(a) $Q_2 = \frac{1}{2} Q_1$ (c) $Q_2 = 2Q_1$
(b) $Q_2 = Q_1$ (d) $Q_2 = 10Q_1$

18. Which type of heat loss between the absorber plate and the adjacent cover plate can be neglected?

(a) Conduction   (c) Insulation
(b) Convection   (d) Radiation

19. Under the equilibrium conditions, what can you say about T as compared to T_1?

(a) T is almost equal to T_1 but is slightly lower.
(b) T and T_1 are exactly equal.
(c) T is considerably less than T_1.
(d) T is greater than T_1.

20. Suppose that the absorptivity of the black absorber plate was normally about 0.90, and we replaced it with a black absorber plate with $\alpha = 0.94$. Over an entire heating season the performance of the solar collector will:

(a) Decrease by 4%.   (c) Increase by roughly 4%.
(b) Stay the same.   (d) Increase by roughly 2%.

21. Suppose that a 15-km/hr wind is blowing. The dominant heat loss mechanism (or mechanisms) for heat transfer between the two glazings is:

(a) Convection and radiation.
(b) Convection and conduction.
(c) Convection.
(d) Radiation.

22. Suppose that a 15-km/hr wind is blowing. The dominant heat loss mechanism (or mechanisms) for heat transfer between the top glazing and the outside world is:

(a) Convection and radiation.
(b) Convection and conduction.
(c) Convection.
(d) Radiation.

23. The heat flow Q_1 from the absorber plate can be written as the sum of two parts. One is the radiative heat loss Q_{rad}, and the other is the convective heat transport Q_{conv} between the two surfaces. Write down the quantitative relation expressing the heat loss Q_{rad}.

24. Suppose that the net heat transported into the cover plate is larger than the heat transported out. What will happen to the temperature of the cover plate?

25. At equilibrium the temperature T_2 of the top cover plate is intermediate between that of the absorber plate and the ambient temperature T_a of the outside world. Explain.

TYPES OF FLAT-PLATE COLLECTORS

The figures below show three types of solar collectors along with their corresponding efficiency curves. The following 10 questions refer to these curves.

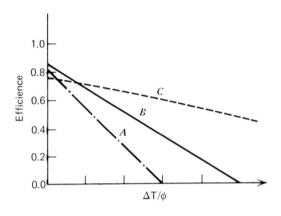

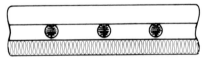

Trickle-type collector

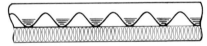

Standard collector

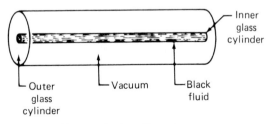

Evacuated tube collector

26. Which efficiency curve corresponds to the evacuated tube collector?

27. Which efficiency curve represents the performance of the standard collector?

28. Which efficiency curve represents the performance of the trickle-type collector?

29. What is meant by ΔT?

30. What is the efficiency of the solar collector represented by curve A when ΔT is very small and the incident solar intensity is large?

31. Which collector is the least expensive to construct?

32. Which collector is the most expensive to construct?

33. Which collector would you utilize if you wished to obtain high-temperature water?

34. Why are three glazings almost never used in a flat-plate collector?

35. The vacuum in the evacuated-tube collector eliminates which type of heat loss mechanism?

The next four questions refer to the picture of one type of selective surface (shown below).

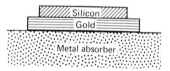

36. The incident solar radiation is absorbed by:
    (a) The layer of silicon.
    (b) The layer of gold.
    (c) The metal absorber plate.
    (d) The water in the pipes in contact with the absorber plate.

37. If infrared radiation were incident on the system it would be reflected by:
    (a) The layer of silicon.
    (b) The layer of gold.
    (c) The metal absorber plate.
    (d) The water in the pipes in contact with the absorber plate.

38. The long-wavelength radiation emitted by the hot absorber plate:
    (a) Passes through the gold and silicon without any appreciable loss.

(b) Is mostly reflected back by the layer of silicon.
(c) Is mostly reflected back by the layer of gold.
(d) Is absorbed by the layer of silicon.

39. The overall emissivity for long-wavelength radiation of this selective surface "sandwich" is typically:

(a) 0.95   (c) 0.10
(b) 0.50   (d) 1/0.10

40. Which of the following is a major problem with present commercial selective surfaces?

(a) Their selective emissivity property tends to degrade when operated in the air for a few years.
(b) They are very inexpensive.
(c) They will not adhere properly to the metal absorber plate.
(d) Their absorptance for visible light is too low.

41. Evacuated flat-plate collectors are better than nonevacuated ones at high operating temperatures because:

(a) Convective heat losses between the two surfaces containing the vacuum are eliminated.
(b) Conductive heat losses from the absorber plate are greatly reduced.
(c) Radiative heat losses from the absorber plate are greatly reduced.
(d) Convective heat losses from the top cover plate are greatly reduced.

42. In the trickle-type flat-plate collector the heat exchange fluid:

(a) Is air.
(b) Flows through pipes to contact with the absorber plate.
(c) Flows under gravity down depressions (corrugations) in the absorber plate.
(d) Is propylene glycol.

43. A severe disadvantage of using a mixture of ethylene glycol and water as the heat-exchange fluid for a flat-plate collector is that:

(a) It won't freeze at $-1°C$.
(b) It is toxic.
(c) It has an extremely low specific heat.
(d) Too much pumping power is required.

44. An advantage of using copper for the absorber plate is that:

(a) It has a much higher thermal conductivity than other metals.
(b) It is cheaper than steel or aluminum.

(c) It will not corrode easily so that water can be used as the heat-exchange fluid.

(d) It weighs less than steel or aluminum.

T_a (Ambient temperature)

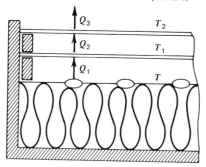

45. The above figure shows a typical cross-sectional view of a liquid-type, solar-thermal collector. The collector is operated in a 15-mph wind. In the following, a number of comments are made about the solar collector. You are to match the comments on the left with the best choice from the list at the right.

_____ 1. The radiative portion of the heat loss Q_1 is

_____ 2. The dominant heat-loss mechanism for Q_3 is

_____ 3. The amount of sunlight passing through the first glass cover is typically

_____ 4. The amount of sunlight absorbed by the black surface is

_____ 5. The peak wavelength of the radiation emitted by the black surface is roughly

_____ 6. The peak wavelength of the incident sunlight is

(a) greater than
(b) 94%
(c) 10,000 nm
(d) 30°
(e) convection
(f) proportional to $T_1 - T_2$
(g) less than
(h) equal to
(i) proportional to $(T - T_1)^2$
(j) proportional to $(T - T_1)$
(k) 0.1
(l) 0.9
(m) 55°
(n) 1000 nm
(o) 500 nm
(p) 84%

_____ 7. At 45°N latitude
the best tilt angle
from the horizon-
tal for the collec-
tor for space heat-
ing applications
is

_____ 8. T_1 is _____ T_2

_____ 9. The far-infrared
emissivity of the
black surface is

46. From the standpoint of heat losses an ideal solar collector would
have the black absorber plate at a constant temperature. In fact,
the absorber plate has a strongly varying temperature distribution.
This latter situation gives rise to a change in the efficiency of the
collector over that for an ideal one under any given operating
conditions. What can be said about the efficiency of the actual
solar collector relative to the ideal one?

47. If a large planar reflector (oriented horizontally) is used in
combination with a flat-plate collector, what is the optimum
collector tilt angle for winter space heating?

48. The efficiency of a solar collector is found to be 43% when the
ambient temperature T_α is 5°C and the input solar radiation is 500
W/m². If the incident solar flux drops to 400 W/m² and the
ambient temperature remains constant, what happens to the
collector efficiency?

The picture below depicts a compound parabolic concentrator. The
following five questions refer to this system.

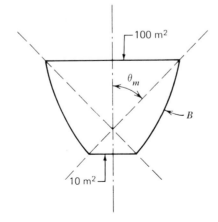

49. What is the popular name of this type of collector?

50. What type of a surface (mathematical name) is B?

51. What is the concentration factor of this system?

52. Is this a focusing-type concentrating system.

53. What can you say about any light that strikes the top surface at an angle of incidence less than θ_m?

9 active space and hot water heating

Solar systems can be classified into two general types: active or passive. Although there is some overlap between the two types, both classifications possess characteristics that are distinctly different from each other. An active system is one in which nonsolar energy is used to effect the transfer of energy. The collection, storage, and distribution of thermal energy is accomplished by moving various heat-exchange media with the aid of pumps and blowers.

From a general functional standpoint, solar space-heating systems are designed to operate in four basic modes. These are (1) direct space heating from the collector, (2) transfer of heat from the collector to the storage system, (3) space heating of the house from the storage system, and (4) space heating of the house from the auxiliary backup system. In addition, the solar system is often designed to preheat water from the incoming water supply prior to its passage through a conventional water heater. The domestic hot water preheat system can be combined with the solar-heating system or can be installed separately. It is not necessary that all systems perform each of the above functions.

The principle elements of a complete active solar-heating system are illustrated in Figure 9-1. The solar energy is captured using some type of a flat-plate collector operating at rather low temperatures. In the collector, either a liquid solution, or air, is circulated to transfer the heat from the collector to a heat storage facility. The heat storage unit may consist of either a large tank of water, a bin containing many tons

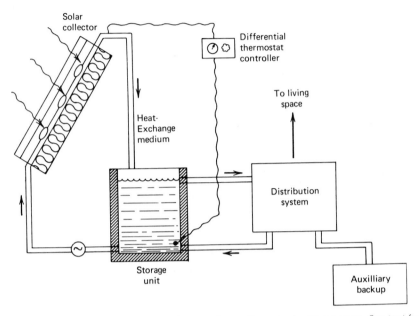

FIGURE 9-1   Schematic drawing of a complete active solar-heating system. See text for more details. Provision for preheating the hot-water supply may also be included.

of small pebbles, or phase-change materials in containers. Heat is then delivered to the house either directly from the collectors or from the heat storage unit. A system of controls is needed to regulate the delivery of heat to the house and to ensure efficient operation of the solar collector. An auxiliary backup heating system is generally necessary to provide heating during extensive periods of low solar radiation. The solar system can also be adapted to provide air conditioning, if desired. Over the past decade the flat-plate collector industry has undergone a period of steady growth. At first this growth was due to the use of flat-plate collectors for active space-heating systems. However, since 1975, passive solar-heating approaches have grown steadily in popularity. The resultant decrease in flat-plate collectors for this activity has been more than offset by a tremendous increase in solar hot water heating. This latter activity is growing in popularity because of its economic attractiveness. A solar hot water heater can take advantage of the large amount of solar radiation available in the summer rather than just for the winter months as is the case for solar space heating.

A.   GENERAL CONSIDERATIONS

Before going on to a detailed discussion of the key components of active solar systems, a few general comments are in order. Keep in mind that not every building site is suitable for good solar energy collection. It is also imperative to locate the site so that an adjacent building will not appreciably shade the solar house or solar collector. This might mean placing the house as near to the north side of the lot as zoning regulations permit. Deciduous trees planted east and west of south-facing windows are advantageous for providing summer shade, but care should be taken to be sure that they do not block off the winter sun.

Designers of solar houses have concentrated on building energy-efficient structures. Although a well-insulated house is of first priority there are a number of other worthwhile items to consider. Spacious high ceilings of the cathedral type waste heat and should be avoided if possible. Careful window design can also be of great benefit. A normal window area on the south side of the house can easily provide most of the daytime heating requirements of a well-insulated house. Too much glass space can create uncomfortably warm, inside temperatures, even in the middle of winter, unless materials such as masonry are used within the structure, which can absorb and store heat. Windows on the north side should be minimized as they are poor insulators and represent a serious heat loss. Heat loss at night can be reduced in a number of ways. It is preferable to double- or triple-glaze all windows. Small screened openings near the floor and ceiling that can be opened and closed give better ventilation than an open window, since natural convection is assisted. All exterior doors should be insulated as well as possible and should have weather stripping.

House insulation is always a wise investment; insulation is cheap compared to the cost of the solar-heating system necessary to overcome heat losses. Sidewalls should contain at least full width, 9-cm fiber-glass batt insulation. Polystyrene foam sheathing may be used instead of the typical impregnated board on the outside of the walls. Other good construction practices such as weather stripping and caulking should be emphasized. Home insulation is usually specified in terms of R values. In mathematical terms, the R value is the reciprocal of the heat transfer coefficient, $U(R = 1/U)$. For good house insulation one should have as low a heat transfer coefficient as possible, which means a high R value. A few examples will help clarify this discussion. The R value of a 7.5-cm, hardwood door is about 2.7 in English units (the units of U are Btu/hr-ft^2-°F). By contrast, 5 cm of polystyrene insulation has an R value of about 8, and 9 cm of fiberglass insulation has an R value of 11. This latter is a minimum target value for the wall design of a new house. Better yet, going to 5-cm by 15-cm (2 in. × 6 in.) studs will permit increasing the insulation to R-19. A well insulated house will have a ceiling insulation of R-30.

An important factor in designing active systems is the collector area needed. Determining this depends on climatological, architectural, and economic considerations. A common method used to estimate the required collector area involves the concept of degree-days (DD). This is defined as the difference between the inside and outside temperatures of a house. Standard practice uses an indoor temperature of 65°F as the base from which to measure degree-days, since most buildings do not require heat until the outdoor temperature is between 60 and 65°. The outside temperature for a given day is then defined as the arithmetic mean of the minimum and maximum temperatures. This idea is graphically illustrated in Figure 9-2.

The usefulness of this concept comes from the fact that building heat losses can be readily estimated. These heat losses can be expressed as joules per day per degree Celsius of temperature difference (average) between the inside of the building and the outside world. Since English units still dominate the American scene, we will instead work in units of Btu/day-°F. Then, if we know the number of degree-days for a given day, the building heat loss for that day is easily calculated. Similarly, the annual space heating needs of the building can be calculated knowing the annual degree-days for the particular location under consideration. Average degree-day data for various locations can be found in a variety of publications. Some typical data are summarized in Table 9-1.

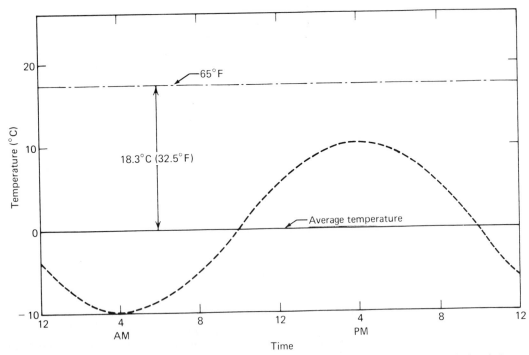

FIGURE 9-2 This figure illustrates the concept of degree-days. The dashed line portrays an idealized picture of the temperature fluctuations throughout the day with an average temperature of 0°C (32°F).

TABLE 9-1
Average Monthly and Yearly Degree-days[a]

SITE	JAN	FEB	MAR	APR	MAY	JUN	JUL	AUG	SEP	OCT	NOV	DEC	TOTAL
New York, NY	1029	935	815	480	167	12	0	0	36	248	564	933	5219
Bismarck, ND	1708	1442	1203	645	329	117	34	28	222	577	1083	1463	8851
Cleveland, OH	1159	1047	918	552	260	66	9	25	105	384	738	1088	6351
Seattle, WN	738	599	577	396	242	117	50	47	129	329	543	657	4424
Caribou, Maine	1690	1470	1308	858	468	183	78	115	336	682	1044	1535	9767
Boston, Mass.	1088	972	846	513	208	36	0	9	60	316	603	983	5634
Omaha, Neb.	1355	1126	939	465	208	42	0	12	105	357	828	1175	6612
Tuscon, Ariz.	471	344	242	75	6	0	0	0	0	25	231	406	1800

[a] All units are degree Fahrenheit days.

EXAMPLE 1

Consider a hypothetical building located in Cleveland, Ohio. Assume that the building heat loss has been found to be 20,000 Btu/DD. What is the space heating requirement for the month of January, and what is the annual space heating load?

For the month of January, the degree-day total is 1159.

$$\left[\begin{array}{c}\text{January space}\\\text{heating load}\end{array}\right] = 20,000 \frac{\text{Btu}}{\text{DD}} \times 1159 \text{ DD} \cong 23.2 \times 10^6 \text{ Btu}$$

Thus, the monthly heating requirement for January is 23 million Btu. For the entire heating season,

$$\left[\begin{array}{c}\text{Annual space}\\\text{heating load}\end{array}\right] = 20,000 \times 6351 \cong 127 \times 10^6 \text{ Btu (or 37,000 kWh)}$$

B. THE SOLAR COLLECTOR

In Chapter 8 the important elements in the detailed design of efficient flat-plate collectors were covered. Now we consider the important practical factors in the choice of the collector to be used in the complete active solar-heating system. These factors include the decision as to whether air or a liquid will be used as the heat-transfer medium, the area of collector needed, the economic cost involved, and whether a commercial or homemade collector is to be installed. Most collectors are located on the roofs of buildings, but consideration should be given to a location at ground level.

If the annual heating requirement of the house is known, then the area of solar collectors needed for space heating can be estimated following the procedure outlined in the previous section. For economic reasons it is not wise to choose a collector area sufficient to provide the entire heating load (including hot water) of the house. A large number of studies have been made to determine the optimum size. The answers vary from one study to the next, depending on such factors as climatic conditions and the projected cost of nonsolar energy. Although it is impossible to give a precise answer to this question, it is safe to say that the solar collector should provide about 50 to 70% of the total annual heating load. The system should not be designed to provide the total heating requirements for the worst winter months of December and January, since then an excess amount of useful heat would be provided in the other months.

The sizing of the solar collector for domestic hot water heating is even simpler, because the economic cost of a few extra square feet of collector is not an important consideration in this case. For a hot water system the costs for installation, storage, and other accessories dominate the total system cost. Because of this, the following rule of thumb will work for most systems: For solar hot water heating use about 65 of solar panels and an 80-gal hot water storage tank.

For the reasons noted above, a small system will not result in much in the way of savings, while a larger system is not practical for most typical households. The responsible family, which makes a moderate adjustment for the weather, will find that this system will provide nearly all the hot water required, except for highly cloudy areas which have long time periods without much sunshine.

About 65 ft^2 of solar panels is close to the optimum for an 80-gal tank. This amount of collector area will capture enough solar energy on a clear day to raise the temperature of 80 gal of water about 150°F.

The solar collector is usually tilted in order to enhance the collection of the direct component of the solar radiation. A rule of thumb estimate for the tilt angle for winter heating optimization is to orient the collector at an angle equal to the latitude plus 10 to 15°. This is born out by the results of a study by the Los Alamos solar energy group, which is shown in Figure 9-3. At a latitude of 42° the optimum collector tilt angle is between 50 and 62°. The calculations were performed for solar systems that could provide either 40 to 74% of the total space heating load. The performance is not very sensitive to the orientation angle, being within 2% of the maximum for tilt angles covering the range of 38 to 65°. It is also of interest to study the azimuthal sensitivity of solar energy collection as presented in Figure 9-4. Although the maximum performance is obtained when the collector is oriented due south, a variation of 30° east or west only reduces the performance by about 2.5

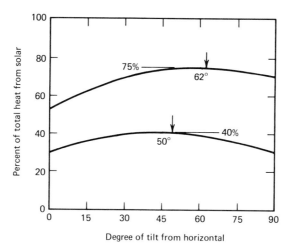

FIGURE 9-3 The amount of useful heat collected by a solar system over the winter heating season in Medford, Oregon (42° North latitude) has been calculated by the Los Alamos solar group as a function of the collector tilt angle. The two curves represent calculations for systems capable of providing 40 and 75% of the total heating load from solar energy when optimally oriented. Small variations from the optimum result in only a very minor reduction in the solar energy collected.

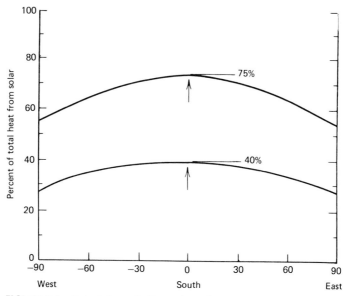

FIGURE 9-4 Calculations similar to those for Figure 9-3 show that a due south orientation is nearly optimum. For exceptions to this rule see the text. The decrease in performance for variations from due south is somewhat greater for colder northern cities than for warmer ones in the southern part of the country.

to 5%. This conclusion is a function of the climatic region, however, as an orientation of 15 to 20° west of south will often be favored for two reasons. First, many areas often experience haze or fog in the early morning hours, which will cut down on the incident solar intensity. Furthermore, higher ambient temperatures in the afternoon will raise the collector efficiency for this period. A good rule for most climatic regions is to orient the collector so as to point directly toward the sun at 1:00 PM (solar time).

The optimum collector tilt angle for domestic hot water (DHW) systems is smaller than the corresponding quantity for space heating. This is because the system is collecting useful energy over the entire year. A standard rule of thumb is to orient the collector (for hot water systems) at a tilt angle from the horizontal equal to the latitude. Analysis has shown that almost the same performance is obtained when the collector tilt angle is 10 to 15° below the latitude angle.

The choice of heat-exchange fluid depends on local climatic conditions, the type of collector chosen, and the individual preference of the user. Air systems have the advantage that they do not freeze up, and no corrosion of the collector is expected. On the other hand, a very airtight circulation system is required, considerable electrical power is needed for moving the air by fans or blowers, and the efficiency of an air-type, flat-plate collector is somewhat below that achieved with liquid systems. Designers of air systems note, however, that the overall efficiency for delivering useful heat to the building is the same for liquid and air systems because of a better extraction efficiency from the heat storage medium when air is used. When a solar collector uses air as the heat-exchange medium, an additional heat exchanger is needed for solar hot water applications.

Solar collectors with a liquid as the heat-exchange medium are most commonly used for hot water applications. These liquid systems present a number of design problems that should be carefully considered. A consideration that is often overlooked is that of providing a uniform flow through the collector. Three common liquid collector configurations are illustrated in Figure 9-5. The serpentine design part (a) is easy to construct but results in a high resistance to the flow of the liquid and a high-temperature rise across the collector. The design shown in part (b) can have problems in balancing the flow of liquid through the parallel tubes. The grid reverse design in part (c) is probably the most efficient. In this case, the flow through the tubes is usually balanced as long as the pressure drop in the headers is less than 10% of the pressure drop across the tubes.

A critical concern with liquid collectors is the possibility of freezing. This is particularly true when ordinary water is used as the heat-exchange fluid. Common solutions are to use an antifreeze solution in a closed-loop system, or to use water in a *draindown* or *drainback*

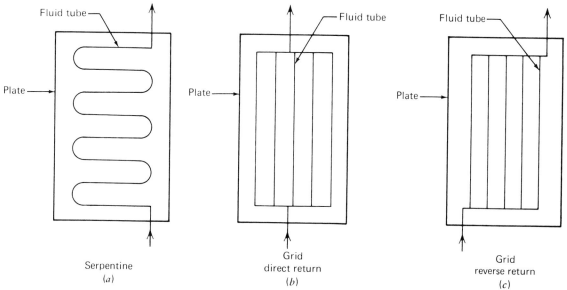

FIGURE 9-5 Three different approaches to distributing the liquid flow across a solar collector.

configuration. More details of these arrangements are presented in a later section of this chapter. Finally, there is the ever-present problem of corrosion. Electrolysis can occur in systems that have dissimilar metals such as copper, aluminum, or steel connected together.

C. STORAGE OF HEAT

All substances, whether they are in a solid, liquid or gaseous phase, are capable of storing heat. The storage of heat can be accomplished either by raising the temperature of the storage medium, in which case it is called sensible heat storage, or by changing the phase of the material, which is referred to as latent heat storage.

Sensible heat is stored (reclaimed) by raising (lowering) the temperature of the storage material during the charge (discharge) cycle. It was pointed out in Chapter 2 that all substances have a specified relationship to water in their ability to absorb heat. This is defined as the specific heat of the material. The heat capacities of several common materials are compared in Figure 9-6. The sloped portions of each curve show the storage of sensible heat.

Latent heat storage is accomplished by changing the physical state of a given material. This is illustrated for water in Figure 9-7. The most

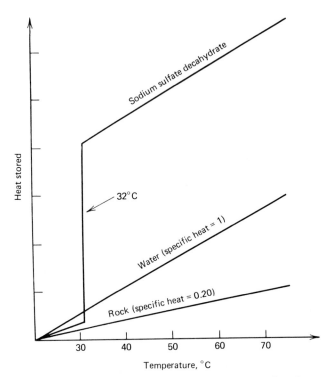

FIGURE 9-6 Comparison of the heat capacities of various storage media. Notice the large amount of heat that can be stored in the heat-of-fusion salt.

suitable phase change is the solid-liquid transition in which freezing and melting of the storage material is involved. Heat storage through phase change has the advantage of compactness, because the heat of fusion of most materials is very much larger than their specific heat. For example, in the case of water the latent heat of fusion is 80 times the specific heat. This means that 80 times as much energy is required to melt one gram of ice as is required to raise the temperature by one degree Celsius. Various types of salts are the most common type of heat-of-fusion materials. Problems associated with this heat storage approach are discussed below.

Storage of sensible solar heat is commonly accomplished with gravel or water. The choice of which storage material to use depends on the type of solar collector chosen. Water is the natural choice for systems employing liquid heat exchangers, while gravel is usually chosen for air collector systems.

The heat storage properties of some common materials are shown in Table 9-2. The heat storage ability of a substance is measured by its specific heat, which was defined in Chapter 2 as the amount of heat

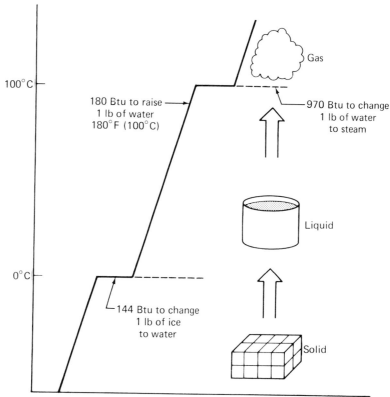

100°C

180 Btu to raise
1 lb of water
180°F (100°C)

970 Btu to change
1 lb of water
to steam

Gas

Liquid

0°C

144 Btu to change
1 lb of ice
to water

Solid

FIGURE 9-7   This schematic drawing illustrates the change of phase of water from solid to liquid and from liquid to a gas.

required to raise the temperature of 1 g of a substance by 1°C. Water can store more heat than most materials because of its larger specific heat capacity. While water has a specific heat of 4.184 J/g-°C, the specific heat of rock is only about 0.84 J/g-°C, a value that is typical of most other solid materials such as iron and aluminum. The common

TABLE 9-2
Heat Storage Properties of Common Materials

MEDIUM	SPECIFIC HEAT	
	J/g-°C	J/m³-°C
Water	4.184	4.184×10^6
Water and ethylene glycol (50–50 mix)	3.35	3.65×10^6
Gravel	0.71	1.16×10^6
Sodium sulfate decahydrate	209 J/g	347×10^6 cal/m³

antifreeze, ethylene glycol, has a specific heat of 2.51 J/g-°C for the pure substance. For the 50-50 mixture of water and ethylene glycol commonly used in solar systems, the specific heat is about 3.35 J/g-°C. For a substance that changes phase the important parameter is the amount of heat required per kilogram (or per unit volume) to cause the change from a solid to a liquid at the melting point of the material. The last item in Table 9-2 is for a typical heat-of-fusion salt, sodium sulfate decahydrate, which changes phase from a solid to a liquid at 32°C.

Water is the most commonly used storage medium, since it is the practical choice when a fluid is used as the heat-exchange medium for the solar collector. Let us illustrate its use as a storage medium by calculating the amount of useful heat that can be stored in an 11,370-liter tank (3000 gal) that contains water at a uniform temperature of 54°C (130°F). Assume that useful heat can be withdrawn down to about 27°C (80°F), giving a temperature rise of 27°C in the storage medium. The heat, Q_3, stored in the storage medium is

$$Q_3 = 11,370 \text{ liters} \times 10^3 \text{ g/liter} \times 4.184 \text{ J/g-°C} \times 27°C$$
$$= 1.28 \times 10^9 \text{ J}$$

This is the equivalent of two to three days of the winter heating load in a mild climate.

The advantage of storing heat with phase-change materials is illustrated in Figure 9-6. A large amount of heat per unit weight is required to effect the melting of a solid as compared to that needed to raise its temperature 1°C. Once sufficient heat is added to cause the entire storage medium to melt, further heat is stored as sensible heat just as is the case for other materials. In principle, heat-of-fusion salts can be used to store large amounts of heat in small volumes of material. In application, the design of phase-change containers is complicated by the need for a large heat transfer surface. Because of this, much of the space-saving features of using a phase-change material is lost since the container needs to be enlarged to obtain the required surface area.

Commercial development of this storage approach has been slow because of problems arising from the behavior of these materials after having been thermally cycled a number of times. Sometimes a portion of this phase-change material sinks when frozen, and it is not a simple matter to return to a homogeneous fluid again upon melting. In addition, phase-change materials have a tendency to supercool, whereby the storage material will fail to solidify at the freezing point. Another problem is associated with the failure of containers after prolonged use. A current approach to overcome some of the difficulties has been to use long, 2.5-cm-thick trays to store the phase-change materials. Much progress has also been made in developing chemicals to overcome the problems of supercooling and segregation. Because of the attractiveness of storing a large amount of energy in a small volume, it is certain that research and development in this area will continue until reliable, economical systems have been achieved.

The size of the solar collector is also influenced by the size of its associated storage system. Since space heating tends to be out of phase with the sun, both daily and annually, storage of the solar energy is required, with the amount again dictated by local conditions. In many areas, one to three days storage is sufficient, whereas in cloudy areas greater storage capacity may be warranted. Theoretically, sufficient storage capacity could be provided to cover the accumulation of fall heat for winter use. For climatic areas where cooling is the dominant requirement the amount of storage capacity may be greatly reduced, since the cooling need is essentially in phase with the sun.

The physical size of the storage material for a space heating system must be chosen with some care. (The size of water tank for domestic hot water applications was dealt with in the previous section.) If the storage unit is too large, it may take days to heat the storage material to a useful temperature. Conversely, a storage unit that is too small might become overheated. This would result in an excessive collector operating temperature and the collector's efficiency would be quite low.

Finally, it is very important to take account of the economic factors involved. The amount of heat supplied by a solar space heating system is shown in Figure 9-8 as a function of the storage capacity. Without any storage capacity, the solar system would contribute only about 20 to 30% of the building heating needs. The solar contribution increases rapidly with storage size until a point near the maximum (say 80% in this example) is reached. Beyond this, the additional solar contribution obtained by increasing the storage capacity is quite small. The most economic point of operation is near the elbow in Figure 9-8.

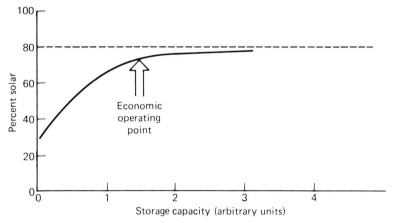

FIGURE 9-8 The amount of solar heat delivered to a typical building is plotted as a function of storage size.

D.   ACTIVE SPACE HEATING SYSTEMS

Although domestic hot water heating is the most important application for active systems, there is sufficient use of space heating systems to justify a brief discussion. Furthermore, a large part of the material covered in this section will be relevant to the hot water applications discussed in the following section.

Liquid solar-heating systems operate in much the same manner as air solar systems so we consider them first. The most important difference is that many liquid systems utilize a heat exchanger in a closed-loop mode of operation. Also, space heating directly from the collectors is usually not feasible with liquid systems, whereas this is part of the normal operating procedure for air systems.

A system for heat storage using water as both the heat-exchange fluid and the storage medium is illustrated in Figure 9-9. Water is drawn from the bottom of the storage tank, pumped through the collector, and returned to the top of the storage tank. When the system is not running,

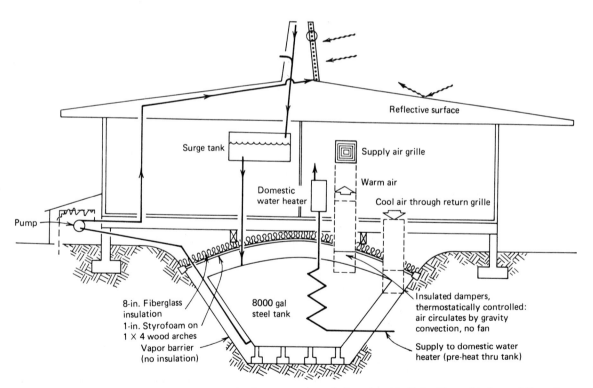

FIGURE 9-9   Schematic arrangement of the solar-heating system designed and built by Henry Mathew of Coos Bay, Oregon. The solar collector is oriented at 82° to the plane of the horizontal, and the reflector is oriented 8° below the horizontal. Air circulation around the storage tank and through the house is due to the force of gravity only. Preheating of the hot water is obtained by passing it through the hot storage tank as shown.

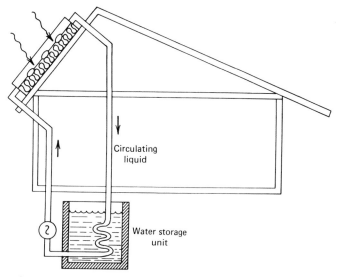

FIGURE 9-10   Often a heat transfer liquid other than water is used to transfer heat from the solar collector to the water storage unit. In this case some type of heat exchange, perhaps with a coil, is needed to transfer the heat from the circulating liquid to the water storage.

the water drains back either into the storage tank or into the surge tank. In the case of the Coos Bay solar house, heat for the house is obtained by having air pass over the warm storage tank. More commonly, the storage tank is insulated and heat is extracted through some type of heat-exchange process.

A closed-loop solar-heating system is illustrated in Figure 9-10. Freeze protection is often obtained by using an antifreeze solution as the heat-exchange medium. Freeze protection for active solar systems is discussed in more detail later. Some disadvantages of water storage systems include the possibility of leaks, the cost of the tank, and the rather large space and weight requirements.

One way to extract heat from the water storage system for delivery to the house is to supply the house with hot water from the top of the storage tank and to withdraw fluid from the cooler bottom for delivery to the collector. This results in the warmest fluid being used for space heating while the collector is supplied with the coolest fluid, resulting in maximum collector efficiency. Water from the storage system can also be circulated directly to the building and passed through coils as illustrated in Figure 9-11.

Key components of any active system are the devices used to sense the system temperature. These are needed in order to activate the collector system when useful solar energy can be collected and to signal the delivery system whenever heat is needed in the building. For

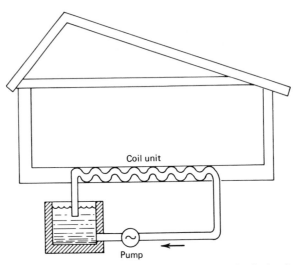

FIGURE 9-11   Schematic illustration of the method of radiant heat distribution using a coil unit in the floor. The coils could also be located in the walls or ceiling of the building.

liquid systems the location of the storage tank sensor is important. This sensor should not be located too near the point where the solar-heated liquid is supplied to the tank. Otherwise, the sensor may indicate that the collectors are not substantially warmer than storage, and pump cycling and shutoff can occur.

Control of the solar collector system is maintained using a differential controller. This device senses water (or air) temperature in various parts of the system to initiate control action. The typical differential controller for liquid space heating and hot water solar system senses the temperature difference between the absorber plate of the solar panel and the bottom of the storage tank. When this difference is more than the critical setting ΔT_{on} (usually 10 to 20°F), the collector system pump is turned on as shown in Figure 9-12. When the solar input decreases to a low level, the temperature difference between the solar panel and the storage tank will reach the critical difference ΔT_{off} (usually 3 to 5°F), at which time the collector pump is turned off by the controller.

Air solar systems differ slightly in design and operation from their liquid counterparts. An important difference is in the storage medium. Crushed rock or gravel is commonly used when air is the heat transfer medium. The heat capacity per unit weight of rock or gravel is only one-fifth that of water, and the heat capacity per unit volume of rock or gravel is about one-third that of water (see Table 9-2). Thus considerably more rock is required to store the same amount of heat stored in a given amount of water. In the earlier example where 1.28 billion J of sensible heat (the energy stored by raising the temperature of a substance) was stored by raising the temperature of 11,370 liters of water 27°C, storage of the same amount of heat by rocks would require three times this volume.

When rock is used for heat storage, only small rocks about the size of a fist or smaller should be used. This is necessary to increase the amount of surface area of the rock. The ability of rock to exchange heat with the circulating air is a function of its surface area per unit volume. Thus, a cubic foot of small rocks will have much better heat transfer properties than a solid rock of the same size. In packing the smaller rocks randomly into the cubic foot of space, only about one-half as much material by weight will be contained compared to the situation with a cubic foot of solid rock. This means the total heat storage capacity is reduced by this factor of one-half, but the large gain in the ability to exchange heat with the circulating air is the most important consideration.

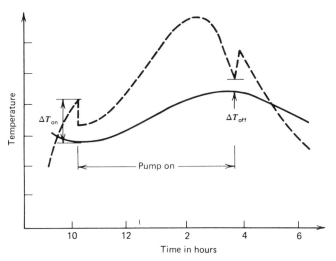

FIGURE 9-12 Illustration of how the differential thermostat controller is used to turn the solar collector pump on and off (for a liquid system). The solid line depicts the temperatures at the bottom of the storage tank, while the dashed line shows the temperature at the collector outlet. For liquid systems, ΔT_{on} is about 6°C and ΔT_{off} is about 2°C.

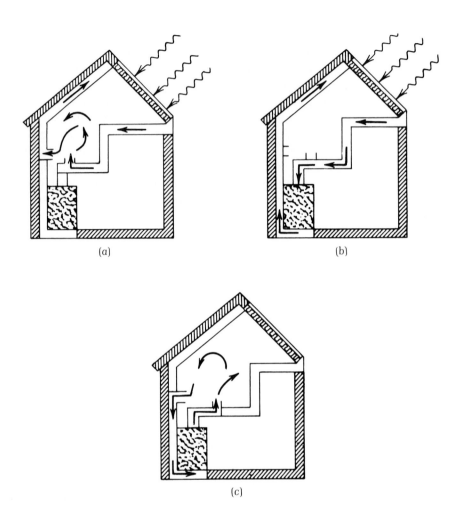

FIGURE 9-13 Illustration of the operating principles of an air-type solar system. *(a)* The heat collected by the solar collector is transferred directly to the building. *(b)* The collected heat is stored in the rock bin. *(c)* Shows how heat is removed from the rock storage and transferred to the building.

A practical solar air system with rockbed heat storage is shown in Figure 9-13. The rocks are contained in a rectangular enclosure, and their weight is supported by a steel grill at the bottom of the container. The central pipe allows the rocks to be shunted out, when the heat is needed, directly to the structure. Rock piles have the advantage that they can neither freeze nor leak. An important aspect of rock storage is the stratification of temperature that is formed. The heated air coming from the collector gives up heat as it passes through the rock storage, exiting as cool air. This results in the entrance portion of the storage being at a higher temperature than the exit portion. In the evening, when the collector is off and the building needs heat, the direction of air flow through the rocks is reversed so that the air leaves from the hottest part of the rock storage system. This results in air to the building that is considerably warmer than if the rock storage were at a uniform temperature. The net result of this temperature stratification is an improved efficiency for the transfer of heat from the storage system to the residence over that obtained with liquid solar systems. This explains the comment made earlier that the overall efficiencies of air and liquid systems are about the same even though the collector efficiency for the liquid system is higher.

The overall arrangement of an air-type solar-heating system is illustrated in Figure 9-14. For an air-type solar collector, the key controls are the house thermostats and the differential thermostat controller. When the house thermostat requires heat, it directs the warm air into the house heating system. The differential thermostat

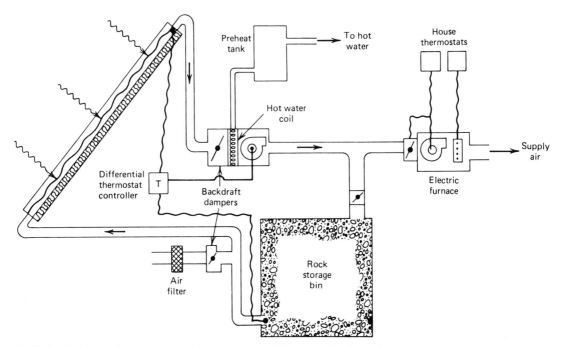

FIGURE 9-14 Possible arrangement for solar space and hot water heating using an air collector and a rock-storage unit. Additional controls, not shown, are needed for the hot water, preheating system. (A design similar to this has been used by Steve Baker.)

controller regulates the flow of air to the collector and back to the storage tank. Temperature-sensing devices are used at the output of the solar collector as well as at the cool end of the storage tank. The fan that circulates air through the solar collector is turned on when the temperature at the collector outlet is somewhat greater than that of the temperature of the cool part of the tank. The operation of the differential control system will result in a pattern much like that shown in Figure 9-12 for a collector using a liquid heat transfer medium. Once heat has been collected and stored it must still be delivered to the building in order to be useful. The delivery system must extract heat from the storage system and deliver it to the house. The controls for this can be thermostats operating in much the same way as a conventional unit. There will be some differences, however. For example, in the case of the air systems shown in Figures 9-13 and 9-14, it is possible to deliver heat directly to the house from the solar collector. At other times the control system will switch the arrangement to permit the flow of warm air from the storage unit to the building. The control system allows returning air from the building to go directly to the solar collector during the day (unless cloudy) and to go to the cool end of the rock-storage system at night.

In the cooler portions of the United States a full-capacity auxiliary heating unit must be provided to supplement the solar-heating system when no useful solar energy is being collected and when all the stored heat has been consumed. The auxiliary furnace is usually arranged to heat the heat transfer medium and distribute it through the same channels as employed for solar heat. This can be done in a variety of ways. Air from the solar collector can be passed through the furnace, if necessary, before circulation to the rooms. Sometimes auxiliary heat is provided to the heat-storage system. For water systems another approach is to heat water in a small, auxiliary water tank, from which it is directed either to the house or to the air-flow heat transfer system.

Use of an adequate number of sensors should be standard design practice with all solar-heating systems. If this is not done it is difficult to evaluate the system performance and to troubleshoot the system in case of problems. Visual sensors for this purpose are highly recommended.

E. SOLAR HOT WATER

The most popular use of active solar systems is to provide hot water. The reason for this popularity is that a large amount of energy can be obtained for a very reasonable cost. The plumbing details, often obnoxious for large active space heating systems, are minimized in the smaller hot water systems. In 1980, only 7% of the active installations were for space heating; the rest were for various applications involving the heating of water. In this section we consider this solar application, concentrating mainly on the special designs used to provide freeze protection. Most of the other features of common solar hot water systems have been covered in the previous section.

In the early 1930s in the United States, widespread commercial use of solar energy for hot water heating was made in Florida and California. The collectors usually consisted of a single glass cover plate mounted over a metal box containing blackened copper tubes. Tens of thousands of these heaters were sold in both states until about 1960. With the onset of cheap natural gas, the solar water heating industry in the United States declined rapidly. By 1973, the industry had almost vanished; only to see a sudden resurgence of interest as the price of fossil fuels started to climb. In Australia, Japan, and Israel, the production of solar water heaters is widespread. In 1960, more than 200,000 solar water heaters were reported to be in use in Japan, with almost the same numbers in domestic installations in Israel and Australia. The reason for the success of solar water heaters in Japan is not only due to the shortage of conventional energy sources, but also to their willingness to adapt their life-style to the solar supply—showering at night instead of in the morning.

The flat-plate collector area needed for solar hot water heating will vary considerably depending on local conditions. A rule-of-thumb estimate is that 1 m^2 of collector area is required to heat 61 liters of water per day. Thus, if a family of five uses 380 liters/day, the necessary area of flat-plate collector would be about 6 m^2 (roughly 60 ft^2). It was noted earlier that a very simple design procedure, good for most solar hot water systems, is to use a solar panel area of about 6 to 10 m^2 (60–100 ft^2) along with an 80- (or 120) gal tank.

When choosing a solar domestic hot water system pay careful attention to the local climatic conditions. In warm climates where freezing rarely occurs, commercial thermosiphoning systems are quite successful. This approach will be covered later, but first we discuss the more common situation for cold climatic regions in which protection against freezing must be provided. The two most common approaches involve either draining the collector or circulating an antifreeze solution.

The need for freeze protection is simply explained. When water freezes, its volume increases by about 10%, giving rise to enough force to break the containing metal pipes. Because of this, freeze protection is a very important system criteria. In fact, active liquid systems are classified according to the freeze protection technique implemented: recirculation, draindown, drainback, and antifreeze systems.

A recirculation system is illustrated in Figure 9-15. In this approach line-pressure potable water is used in the collector circuit and will not

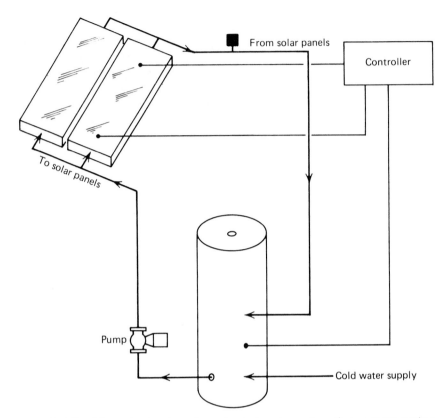

FIGURE 9-15 Schematic drawing of a solar hot water system that protects against freezing by recirculation.

drain when the circulating pump is turned off. When freezing conditions are sensed, water from the storage system is pumped through the collector system. The flow of water at any temperature is usually adequate to prevent freezing under not excessively cold conditions. Freeze-protection sensors are commonly located on the inlet manifold of the first solar panel and on the outlet manifold of the last solar panel. The inlet manifold sensor is set to start recirculation at night and the outlet manifold sensor (at the higher temperature) stops the flow of water.

This approach to freeze protection provides reliable protection for moderately cold climates when electricity is available and the pump is operational. However, if a power failure occurs during freezing conditions, the system may freeze up.

A better approach is to use a draindown or drainback system. The draindown approach is shown in Figure 9-16. When the sensor detects freezing temperatures, the system is allowed to drain through a special valve. The collector loop is then refilled (under line pressure) when the temperature of the freeze-protection sensor rises above the preset value or the differential in temperature between the solar-panel absorber plate and the storage tank is large enough to begin collection. Either one or both of these criteria may be used for refilling.

There are a number of features that are crucial to the operation of a draindown system. Line-pressure water must be stopped from entering the collectors whenever power is removed (freeze conditions sensed or the collector turned off because of low insolation). A normally closed valve must open to allow draining when power is removed. A vacuum breaker and an air vent allow air to enter or leave the collector during draining or refilling, respectively. Draindown systems are prone to

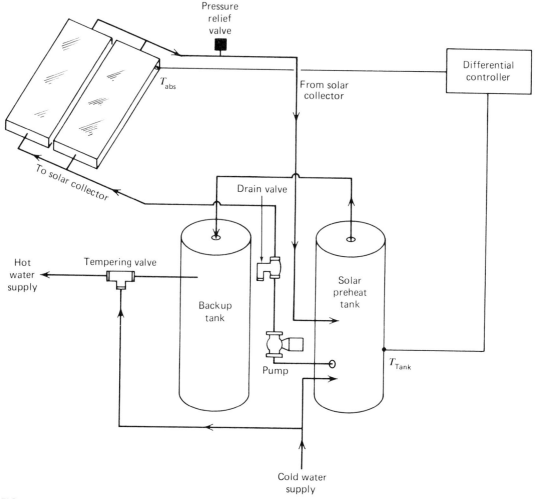

FIGURE 9-16 Schematic drawing of a complete active solar hot water system. The specific design shown here is known as a *draindown* system. See text for details.

corrosion since the piping is often exposed to air and the water cannot be treated with inhibitors. Redundant frost protection in draindown systems sllows them to collect useful energy under conditions in which it is freezing outside, but sufficient solar radiation is available to satisfy ΔT_{on}.

Drainback systems utilize a closed loop with a heat exchanger in the bottom of the storage tank as illustrated in Figure 9-17. When the circulation pump is turned off, the water (usually distilled) in the closed loop drains back into a small reservoir. In this approach water is exposed to the environment only during the solar-collection period. Drainback systems can be used in any climate and do not require a freeze-protection sensor, any special control valves, or electrical power for protection. Draindown and drainback systems need sloped piping to work.

Antifreeze systems use a heat-exchange loop to separate a very low-temperature freezing fluid in the collector loop (which does not drain

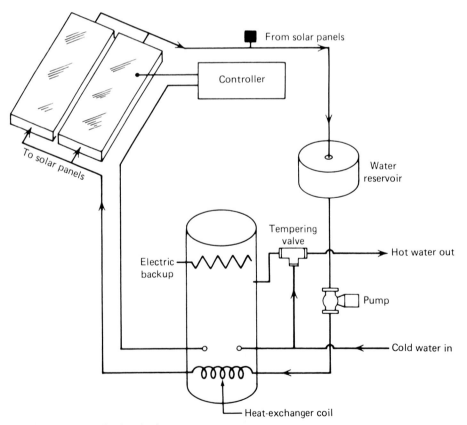

FIGURE 9-17   Drainback solar hot water system.

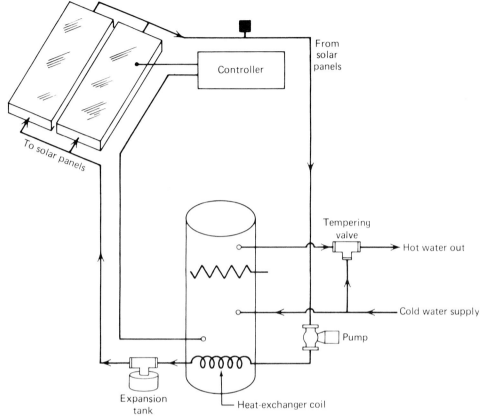

FIGURE 9-18 Schematic drawing of a solar hot water system that protects against freezing by using an antifreeze solution such as etylene glycol.

down) from the potable water supply. This approach is illustrated in Figure 9-18. As with the drainback system, a heat-exchanger loop in the storage tank is needed. Fluids commonly used are water-glycol combinations, silicons, hydrocarbons, and air. Glycol solutions, because of their availability and relatively low cost, have been widely used. Problems are the need for double-wall heat exchangers to protect against toxic fluid contamination and breakdown of the glycol solution to form acids that can corrode system pipings. Silicons and hydrocarbons have poor thermal characteristics that reduce collector efficiency.

Passive (or thermosiphon) systems do not require power for freeze protection or operation. This system is illustrated in Figure 9-19. In this approach a tilted collector is connected to a separate, well-insulated water storage tank by insulated pipes. The bottom of the storage tank is a little over one-half meter higher than the top of the flat-plate collector, with circulation occurring through thermosiphoning. When the water

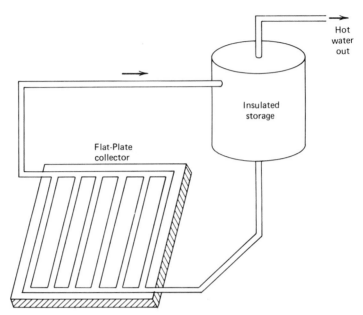

FIGURE 9-19 Illustration of a typical solar hot water heating system operating on the thermosiphon principle. See text for details about freeze protection, but note that this passive approach is best for climatic areas where freezing is not a severe problem.

in the collector is heated by the sun, it becomes less dense and rises up to the top of the storage tank. The warm water is replaced by the more dense, cool water coming from the bottom of the storage tank. As long as the sun shines, the water will circulate, and its temperature will rise. Flow in the reverse direction is generally avoided because of the height differential between the inlet and outlet ports. Some systems have a check valve that allows water flow in only one direction.

Piping from the collector to the storage tank should be sufficiently large so that thermosiphoning will occur without the need for extreme temperature differences. Typically, a 2.5-cm pipe suffices. The storage tank is commonly large enough to hold a two-day supply of hot water. Standard practice indicates that the average person consumes about 75 to 115 liters per day. A larger storage tank will carry over longer sunless periods, but will cost more and will require a larger collector area. The supply pipe to the bottom of the collector should feed from the bottom (coolest part) of the storage tank. Hot water from the collector should feed almost to the top of the storage tank. The inlet should be sufficiently below the top so that about a half day's worth of hot water is above the inlet. This is done because water coming from the collector may not always be as hot as that at the top of the storage tank because of low-insolation rates.

Freeze protection for thermosiphon systems is obtained by using electric-resistance heaters, manual draindown, automatic draindown, or an antifreeze solution. The electric-resistance approach is popular but suffers from the possibility of freezing during a power failure. Manual draining requires an energetic, well-informed owner who will drain the system whenever the possibility of freezing is imminent.

When the space-heating requirements of a home are provided for by an active solar system, the large heat-storage system can be used as a preheater for a conventional water heater. For most air or water space heating systems, a simple heat exchanger transfers heat from the storage to the water on its way to a regular hot water heater.

F. SOLAR COOLING

The use of solar energy to provide cooling is attractive because the cooling is needed when the solar energy is most available. If the local climate requires both heating and air conditioning, a combined system to provide for both of these building needs may prove particularly economical because of a better overall system-usage factor. Although at present only a small fraction of the energy utilized for space heating and cooling goes to meet the latter need, the number of buildings incorporating air-conditioning units has been steadily increasing. From the point of view of conservation of energy it makes good sense to use the energy incident on a structure to provide the required cooling.

In this section we discuss various methods of obtaining solar-assisted cooling. Before doing this it is worthwhile to point out a few general conclusions relevant to the solar collectors used for combined systems, providing both heating and air conditioning. First, the optimum collector tilt angle is nearly equal to the latitude angle. Also, the optimum collector area needed is always larger than the area that would be used by a system providing space heating only. Finally, the solar collector should operate at temperatures considerably higher than those required for heating. These higher operating temperatures imply the use of low-heat-loss collectors. One way to achieve this is with evacuated collectors.

Schemes to provide solar cooling can be divided into three broad categories:

1. **Natural Cooling Methods.**   Heat is given off to the night air and sky by convection and radiation. The storage unit is utilized during the day for absorption of heat from the living space and the stored heat given off at night.

2. **Dehumidification Systems.**   Often the removal of moisture from the air in the living space is desirable. This approach is typified by

desiccant systems in which air is dried and then cooled by evaporation. Solar energy is used as the source of heat for regenerating the air-dehumidifying agent.

3. **Systems That Provide Refrigeration.**   The cooling can be provided by a system using absorption refrigeration or by a conventional vapor-compression, air-conditioning unit where the needed mechanical energy is derived from the solar input.

Given suitable climatic conditions, natural cooling methods have been used with some success. This approach, graphically illustrated in Figure 9-20 takes advantage of the radiation from hot solar collectors into the night sky. In effect, the typical active solar-heating system is operated in reverse. Heat is transferred during the day from the warm house to the storage medium. Air or water can then be circulated from the storage unit through a large radiating surface, dissipating heat to the sky. This approach works best in the southwestern region of the United States, where there are large differences between day and night temperatures.

Evaporative cooling is obtained through the loss of heat that occurs

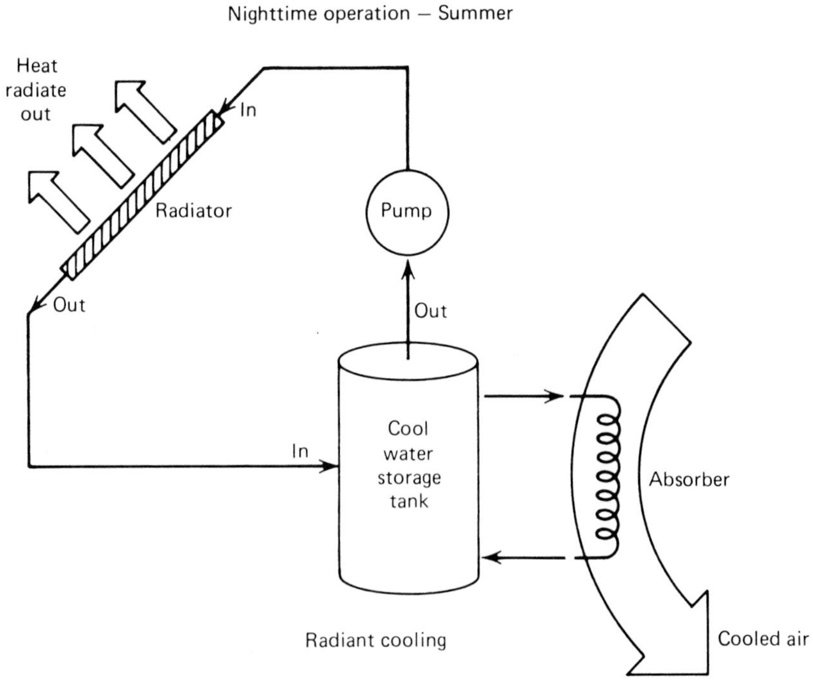

FIGURE 9-20   Schematic illustration of natural cooling achieved by operating a typical active solar/system in reverse.

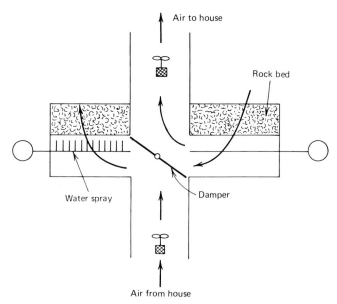

FIGURE 9-21 The Australian rockbed regenerator system provides cooling using the evaporation of water. (Courtesy of J. I. Yellot, "Solar Radiation and its Uses on Earth," published in *Energy Primer*, pp. 4–24, Portola Institute, Menlo Park, Calif.)

when moisture is evaporated into the air. The process of passing air through pads that are saturated with water and evaporating some of the moisture picked up by the air has been used for many years as a means of reducing the temperature of desert air to a tolerable level. The problem here is the subsequent rise in relative humidity.

A modification of the simple evaporative cooler has been developed in Australia and given the name, "Rock Bed Regenerative Air Cooler." A schematic example of such a system is shown in Figure 9-21. Two beds of rocks are used, which are separated by an air space in which a damper is located. At the beginning of a cycle, water is sprayed onto one of the rockbeds for 10 to 15 seconds, thoroughly saturating the rocks. The air from the house evaporates the moisture, cooling the rocks on one side. On the other side, cool air is obtained by drawing air into the house over the previously cooled rockbed. Only a very small amount of moisture is added to this air, since it is the rocks that have been cooled and the only moisture remaining is that small amount adhering to the rocks' surface. The advantage of this technique is that the evaporation of 1.5 gal of water is equivalent to about 1 ton of refrigeration. The amount of power needed is only about 10% that of a normal refrigeration system. (Unfortunately, this system will not work when the outside humidity is high.)

The most common way to achieve solar cooling is to use absorption refrigeration. Thus far, only absorption cycles using lithium bromide and water, and ammonia and water have obtained any commercial success. The arrangement of a typical absorption cooling unit is shown in Figure 9-22. In this approach pressurization is accomplished by dissolving the refrigerant in a liquid in the absorber section. The solution is then pumped to a higher pressure. The low-boiling-point refrigerant is driven from the solution by adding solar heat in the generator and the vapor passes on to the condenser where heat is given off. The condensed refrigerant passes on to the evaporator through the expansion valve. In the process of evaporation, heat is extracted from the air to provide the cool air (or liquid in some cases) needed for cooling. Finally, the refrigerant vapor passes back to the absorber where it is absorbed back into the solution.

In the ammonia-water, heat-actuated refrigeration system, water is the absorber. At low temperatures water can absorb a large quantity of ammonia. When heated at high pressure the ammonia is given off as a gas. The ammonia can then be condensed back to liquid form and allowed to expand through a valve. Some of the ammonia will evaporate and cool down the remainder. The only external mechanical

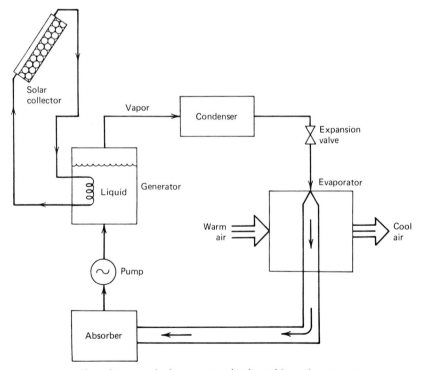

FIGURE 9-22 Flow diagram of a heat-actuated solar refrigeration system.

work needed is that required to raise the pressure of the ammonia-water system. The chilled water resulting from the process is used to provide the desired air conditioning. The lithium-bromide system uses water as the refrigerant. Lithium-bromide will absorb large quantities of water when cool and give it up as water vapor when heated. Regardless of the type of heat-actuated refrigeration process used, operating temperature is a trade-off when coupling a solar collector to a heat-actuated refrigeration unit. This is due to the fact that the efficiency of the solar collector decreases with increasing collector temperature, while the performance of the refrigeration unit increases with generator temperature. One commercial lithium-bromide system delivers four times the amount of cooling at 104°C than it does at 82°C. One of the key problems associated with absorption cooling systems is associated with the start-up transient. Cooling does not begin immediately when the unit is turned on, but instead starts only after the generator has reached a certain minimum temperature. In practical systems a certain amount of time is required to heat up the generator to the point of maximum efficiency. For systems in which the cooling unit is alternately on and off, this time delay results in an average performance considerably below the optimum.

Considerable attention has been given to methods whereby thermal energy obtained from the sun would be converted to some form of mechanical output. This in turn would be used to drive a more conventional vapor-compression refrigeration unit, which then performs the function of cooling the building. Compare this method with the absorption equipment described earlier where the thermal-energy input directly produces the cooling function. Determining which of these cooling approaches will be best for various regions of the country must await further technological development.

Systems in which moisture is removed from the air through the use of desiccants are most suitable for air systems. In this approach, air-drying agents that can be regenerated by heat are used. Their usefulness is primarily for hot, humid climates where a large portion of the cooling load is for the removal of moisture from the air. The solar energy is then used as a source of heat for the regeneration of the air-dehumidifying agent. One common commercial unit employs triethylene glycol as the dehumidifying agent. Dehumidification is accomplished by alternately absorbing the moisture from the air and then removing it from the resultant solution of desiccant and water by using solar heat.

REFERENCES   1. P. Lunde, *Solar Thermal Engineering: Space Heating and Hot Water Systems*, (New York: Wiley, 1980).

An excellent technical treatment of the topics in this chapter. The interested reader needs a decent mathematical and physical background. The chapter on heat transfer is particularly easy to read and contains a wealth of useful information.

2. M. J. Fisk and H. C. Anderson, *Introduction to Solar Technology*, (Reading, Mass.: Addison Wesley, 1982.)
    Chapters 5 and 6 contain some useful information on active systems.

3. J. I. Yellott, "Solar Radiation and Its Uses on Earth," in *Energy Primer* (Menlo Park, Calif.: Portola Institute, 1974), pp. 4–17.
    An excellent treatment of solar heating and cooling systems at a level that is understandable to the nonspecialist.

4. G. O. G. Löf, "The Heating and Cooling of Buildings with Solar Energy," Chapter 11, in *Introduction to the Utilization of Solar Energy*, A. M. Zarem and D. D. Erway, eds. (New York: McGraw-Hill, 1963).

    Discusses the general aspects of solar heating and cooling in a nontechnical manner. Heat exchange at the collector, types of storage systems, delivery of heat to the building, control systems, backup heating, and solar-cooling possibilities are reviewed as of 1963. Includes a description of several experimental homes.

5. D. Watson, *Designing and Building a Solar House* (Charlotte, Vt.: Garden Way Publishing, 1977).
    Chapters 2, 4, and 5 of this book contain a lot of material relevant to the subject matter of this chapter. The presentation stresses the architectural features of solar heating.

QUESTIONS

1. Suppose that hot water is stored at a temperature that is 20°C higher than room temperature. Since 1 liter of water is about 1000 cm^3, about how many liters of water are needed to store 82 million Joules of heat? (Remember that 1 cm^3 of water weighs about 1 g.)

    (a)   100 liters   (c)   1000 liters
    (b)   200 liters   (d)   2000 liters

2. By roughly what factor would you expect the rate of heat loss of a house to decrease if the thickness of insulation in the walls and ceiling is doubled?

    (a)   None   (c)   4
    (b)   2   (d)   8

3. Consider normal winter operation of a solar collector at a latitude of 45°. The optimum tilt angle (from the horizontal) for solar space heating is near:

    (a)   30°   (c)   60°
    (b)   0°   (d)   90°

4. Suppose that the solar intensity in your climatic area were composed entirely of diffuse radiation. Your latitude is 50°. What would be the optimum tilt angle (from the horizontal) for your collector?

(a) 45°   (c) 60°
(b) 0°   (d) 90°

5. For a 20°C rise in temperature, which of the following materials will store the most heat?

(a) A ton of rocks   (c) A ton of water
(b) A ton of wood   (d) A ton of iron

6. The performance of a heat-actuated, solar-cooling system is most efficient at which of the following average temperatures for the absorber plate? Assume an ambient outside temperature of 25°C.

(a) 15°C   (c) 70°C
(b) 45°C   (d) 100°C

7. In a climate where there is a great deal of morning fog, how should the solar collector be oriented?

(a) Directly south   (c) 30° east of south
(b) 15° east of south   (d) 15° west of south

The figure below illustrates the behavior of four different types of heat-storage materials. The following seven questions refer to this figure.

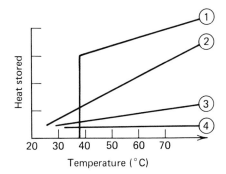

8. Which material has the largest specific heat?

(a) 1   (c) 3
(b) 2   (d) 4

9. Which curve depicts the behavior of a heat-of-fusion salt?

(a) 1   (c) 3
(b) 2   (d) 4

10. Which curve most probably depicts the behavior expected for water storage?

   (a) 1 (c) 3
   (b) 2 (d) 4

11. At what temperature does the phase-change material melt?

12. What is meant by specific heat?

13. If the temperature never rose much above 38°C, which material would you choose to store the greatest amount of heat?

14. What is meant by sensible heat storage?

15. The control system for a conventional solar hot water system is arranged to turn on the fluid in the solar collector when the temperature of the fluid leaving the collector is above (by a preset amount) the temperature of the water at the bottom of the storage tank. This is because:

   (a) This assures a finite temperature drop across the collector.
   (b) This maximizes the efficiency of the solar collector.
   (c) This ensures that the temperature of the fluid flowing into the storage tank will be higher than that of the water in the tank.
   (d) This reduces the collector's upward heat loss.

16. For optimum storage of heat with rocks, small pebbles should be utilized because:

   (a) This gives a large surface area to mass ratio.
   (b) This gives a small surface area to mass ratio.
   (c) This increases their specific heat.
   (d) The economic cost of the rocks is decreased.

17. One of the following statements about using water as the heat-exchange fluid is not true.

   (a) It is convenient to obtain and handle.
   (b) It has a large heat capacity.
   (c) A large amount of electrical power is required to run the water pump.
   (d) Freeze protection can be obtained by mixing with ethylene glycol.

18. In a conventional, warm-water heating system, the auxiliary backup heating system should most certainly not be designed to do one of the following:

   (a) Heat water in the delivery flow system.
   (b) Heat air directly in the air delivery system.

    (c)   Heat water in the storage tank directly.

    (d)   Heat the circulating, solar collector heat-exchange fluid.

19.   Eaves are recommended for south-facing windows because:

    (a)   They provide the necessary winter shade.

    (b)   They cut down on the solar input in the summer.

    (c)   They act as a reflector for winter optimization.

    (d)   They increase the solar input in the summer.

20.   If air is used as the collector heat-exchange medium, what type of material will be used for storage of heat?

21.   A typical differential controller for a water flat-plate collector system turns the collector water pump on when the temperature at the collector is about 10–15°F above the temperature of the water in the bottom of the storage tank. Explain.

22.   Rock-storage systems use "fist-sized" rocks, rather than much larger ones. Explain why this is done.

23.   Name three ways in which a liquid-type flat-plate collector system can be protected against freezing.

24.   Name two disadvantages of active solar systems that use air as the heat-exchange medium rather than water (or some other liquid).

25.   Name two possible advantages of active solar systems that use air as the heat-exchange medium rather than water (or some other liquid).

26.   The amount of sensible heat stored by rock or water storage systems is proportional to two quantities which can be varied. Name them.

27.   What are the three most common (general) types of storage materials used in active solar houses?

28.   Why does a flat-plate collector perform better than a concentrating collector on a cloudy day?

29.   Which stores the most heat, 5 lb of water or 5 lb of rock (at the same temperature difference)?

30.   Name a disadvantage of heat-of-fusion salts.

The figure on the next page depicts one type of solar hot water system. The next three questions refer to this figure.

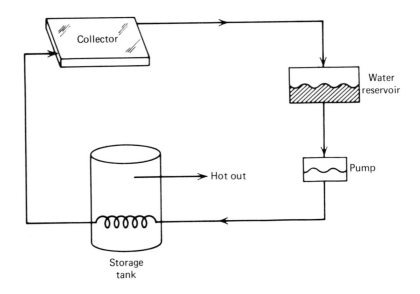

31. How does the freeze protection arrangement work?

32. What is the purpose of the small water reservoir?

33. What is the typical daily need (in gallons of water) of a family of four?

34. What present (1982) federal incentive favors the purchase of a solar hot water system?

The figure on the next page shows another type of solar hot water system. The next four questions refer to this figure.

35. How is freeze protection obtained with this system?

36. What is the popular name of this approach?

37. Explain why a closed-loop system with a heat-exchange coil in the bottom of the tank is not feasible with this approach.

38. Why is water pumped from the bottom of the storage tank to the solar collector rather than from the top of the storage tank?

39. Water changes from the solid stage (ice) to a liquid at zero°C. A large amount of heat can be stored in this way. What is wrong with using ice as a phase-change storage material?

40. In the following problem you are to compare water and air active collector systems. The item on the left describes a quality of one of the two systems. Put a cross in the appropriate box to the right to indicate which of the systems is most appropriate for this quality.

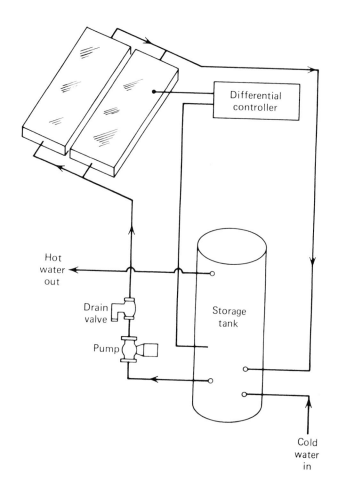

	Air	Water
(a) Uses air as the heat-exchange medium.	☐	☐
(b) Has the highest collector efficiency, under the same operating conditions.	☐	☐
(c) Leakage of the heat-exchange fluid is often difficult to detect.	☐	☐
(d) Has no problem with freezing.	☐	☐
(e) Uses rock as the storage medium.	☐	☐
(f) Uses water as the storage medium.	☐	☐
(g) Commonly deployed in the draindown mode.	☐	☐
(h) Was utilized in the Boulder, Colorado home of G. Löf.	☐	☐
(i) Was utilized in the Washington, D.C. home of H. Thomason.	☐	☐
(j) Requires the most mass for storage.	☐	☐

(k)   This system always uses air as the heat-exchange
      medium for delivery of heat to the house.   ☐   ☐

41.   Assume that a liquid-type, heat-exchange fluid is used in conjunc-
      tion with a water storage tank. Describe two different ways that
      heat can be passed into the house using warm air as the heat
      transfer medium.

42.   Describe two different arrangements for getting heat into the
      house from a water storage system using water as the heat transfer
      medium.

43.   Suppose that you want to obtain approximately equal amounts of
      useful solar energy for space heating in the winter and air
      conditioning in the summer. Would the use of a reflector be a
      good idea? Explain.

44.   8.9 cm of fiberglass insulation in a conventional 5- by 10-cm (2 in.
      × 4 in.) sidewall gives an insulation value of R-11. Also, 14 cm of
      fiberglass insulation in a conventional 5- by 15-cm (2 in. × 6 in.)
      sidewall gives an insulation value of R-17. What R value would be
      obtained if 5- by 20-cm studs were used in the sidewalls of a
      house?

45.   When a homeowner installs a small solar-heating (or cooling)
      system on the house, he or she is in fact creating a small-scale
      power station. Discuss what could be done by local and federal
      government to recognize this fact, allowing the homeowner to be
      economically competitive with other energy sources.

46.   In evaluating the economics of water heating with flat-plate
      collectors, it is often assumed that the averaging process takes
      place over 20 years. However, if the solar system lasts 30 years,
      what does this do to its economic viability.

10 solar thermal power

Utilization of the energy from the sun for water and space heating is now well established. The rapid growth of this part of the solar industry is due to both the solid technical base provided by the early workers in this field and the economic attractiveness of solar applications as the price of fossil fuels has escalated. Future development efforts will probably stress making active systems more reliable and durable and improving the public awareness of passive techniques.

The technological base for solar-electric applications is not as well developed. There are two important approaches to converting solar radiation into useful electric power. The first involves the direct conversion to electric current using some type of photovoltaic device. The other general approach involves the conversion of solar radiation first into heat and subsequently into electrical power utilizing some type of thermodynamic cycle. The development of this latter approach is the subject matter of this chapter.

Solar-thermal power plants using heat engines presently seem most likely to be the first large-scale electricity producers. For small-scale applications solar cells presently appear to be the most attractive possibility. This generalization could easily be proved wrong, of course, if some new development (such as inexpensive, high-efficiency amorphous silicon solar cells) appears on the scene (see Chapter 11).

The technical feasibility of solar-thermal systems was established near the turn of the last century. Some of these pioneering developments, such as the Shuman and Boys solar power plant at Meadi, Egypt, were reviewed in Chapter 4. The numerous solar-thermal power systems that are under development today are based on the same principles as were the early applications. The large improvements being made are due primarily to modern design and manufacturing techniques.

The development of efficient thermal devices to produce mechanical power or electricity usually requires concentrating optical systems. The subsequent increase in technical complexity is more than offset by the higher temperatures available in the absorbing system and, equally important, by the greatly reduced thermal losses. These two topics will be discussed in detail.

It is worthwhile to summarize some of the important concerns about solar-thermal electrical generation schemes. Since the incident energy density is low, concentrating devices will be used, thereby requiring large initial expenditures for equipment. The capital expense for the collecting apparatus and power-conversion devices is the dominant feature in determining the ultimate cost per kilowatt hour of electricity, since the fuel is free. The only real pollution problem to be faced is that of the disposal of waste heat; solar energy is no different from other sources of energy in this respect. On the other hand there is no emission of particulate matter or noxious gases to contend with and no radioactive waste to worry about. Solar power plants will be located at large distances from the major population centers, so that transmission losses and costs must be carefully considered. Perhaps the most troublesome feature of solar-electric plants is associated with fluctuations in the incident solar energy. This can be due to clouds, or to the diurnal variation from day to night. Adequate means of storage must be developed.

A. APPROACHES TO SOLAR ELECTRICITY

There are many different schemes that have been devised to generate power or electricity from solar radiation. In this section a few of the important approaches are reviewed. The different approaches to the generation of power from solar energy can be divided into the following categories:

1. **The Direct Conversion of Solar Radiation into Electricity Using Solar Cells.** Photovoltaic devices are presently economically viable for remote site applications. Their widespread use will depend on the results of the present research and development program. This topic is of sufficient importance that the next chapter is devoted entirely to the subject.

2. **The Central Receiver Concept.**   A heat-exchange fluid located on top of a tall tower is heated by the solar energy reflected from thousands of individual mirrors that track the sun (heliostats). This approach would involve concentrating the incident solar intensity by a factor of 1000 or more. The solar tower facility is illustrated in Figure 10-1. This approach looks very promising and the next section of this chapter is devoted to solar towers.

  A variation on the central tower concept is to absorb solar radiation in air and allow the warm air to rise in a tall tower and run a wind turbine. Thus, a large area of land is covered with plastic and the absorbed solar radiation warms the air under the plastic. The warm air then rises in a tall tower and electricity is generated near the top.

3. **The Conversion of Thermal Energy into Electricity by a Large Number of Distributed Collectors Employing a Moderate Degree of Concentration.**   This approach enhances the incident solar intensity roughly by a factor of 10 to 100 when reflectors are used, making it possible to obtain temperatures over the range of 100 to 500°C. Intermediate-temperature systems have the advantage that they can be fairly small in size and can be located close to the ultimate energy user.

  Two examples of smaller focusing devices are shown in Figure 10-2. In the parabolic trough system, sunlight is focused by a trough-shaped parabolic reflector onto a long, narrow heat absorb-

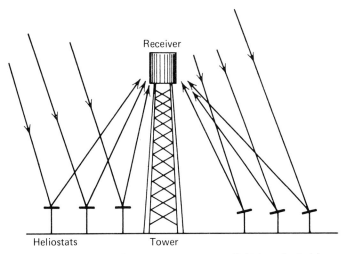

Receiver

Heliostats   Tower

FIGURE 10-1   In the solar tower concept, sunlight is collected by a large field of mirrors that track the sun. The incident solar radiation is reflected and absorbed by a receiver located at the top of a tall tower.

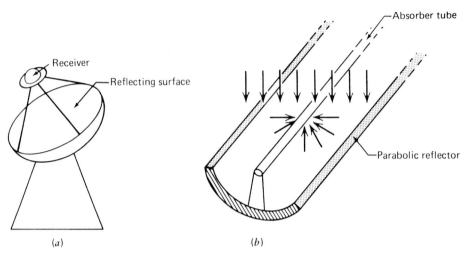

FIGURE 10-2   The two-axis tracking parabolic dish is illustrated schematically in part (*a*). Part (*b*) shows the parabolic trough collector.

er. Another example is the parabolic dish system in which sunlight is focused by a bowl-shaped parabolic reflector onto a small, centrally positioned absorber.

4.   **Generation of Electricity from Biomass.** Photosynthesis is nature's way of converting solar radiation to a useful form. Many different schemes for extracting the energy from biomass are presently under study. This topic is not discussed in this text.

5.   **Ocean Thermal Energy Conversion.**   In this approach power will be generated in a turbine, driven by the small temperature gradients present in the ocean between the surface and the water at depths of hundreds of meters.

6.   **The Solar Salt-Gradient Pond.**   This is a still body of shallow water that collects solar radiation. This collected energy is then stored in a bottom salt-heavy layer of water. By maintaining a gradient of salt in the pond, with the highest salinity near the bottom, heat loss by convection is reduced to an almost insignificant amount. In this way temperatures of the order of 60 to 90°C can be obtained.

B.   SOLAR TOWERS

In the central receiver concept, large sun-tracking mirrors (heliostats) concentrate the solar energy incident on the earth and redirect it to a tower-mounted receiver. The redirected radiation is absorbed by the receiver and converted to heat which is used to vaporize a circulating fluid. The high-temperature, high-pressure gas is returned to the ground and used to power an electric-turbine generator and simultaneously to deliver energy to a thermal storage unit for deferred operation. Currently, the working fluid is water, but other fluids such as gas, molten salt, organic liquids, and liquid sodium are under consideration. The receiver fluid may be passed through a heat exchanger to heat the working fluid or it can be used directly as the working medium in a thermodynamic process.

As an example of the size requirements of a solar tower facility, the mirror field for a 100-MW$_e$ facility would consist of about 24,000 heliostats, each with an area of about 50 m^2. This reflecting array would be deployed over an area of about 5 km^2 (1.8 mi^2) if located in the central California desert area. The tower for a power plant of this size would be 260 m high. The central receiver approach is suited to large-scale (100-MW$_e$) applications but is also adaptable to medium-size applications.

In order to achieve good efficiencies for a thermal power plant, it is necessary to have the source of heat at medium to high temperatures. Since the incident solar radiation has a low energy density, some means of concentrating the incoming solar radiation to a much higher energy density must be devised. The diffuse radiation comes from all directions in the sky, so that only the direct solar beam can be concentrated; solar electric plants will naturally be located in areas where the direct solar radiation is large. The solar tower concept achieves a high-concentration factor through the use of a large field of reflecting mirrors. In this case the solar energy density on the central receiver will be over 1000 times as great as that for the normal incident solar radiation. The temperature achieved at the focus of the mirrors (the receiver) will be of the order of 600 to 1200°C. It is quite difficult to increase the temperature markedly above these values. Increasing the concentration factor from 1000 to 10,000 would only increase the temperature of a system, initially at 1000°C, to about 2000°C.

B.1. Historical Review

The development of large solar furnaces in France under the direction of Felix Trombe was a precursor of the present power tower effort. At the end of World War II, Dr. Trombe interested the French government in building a solar furnace for research at Mt. Louis in the Pyrenees. This was an area with clear air and large annual solar flux. Trombe, who began his research in this area in 1946, was convinced that solar furnaces had enormous possibilities of high-temperature research. His first research equipment consisted of converted German antiaircraft searchlights and parabolic sound collectors, used in wartime for detection of enemy aircraft engines.

The Mt. Louis solar furnace was completed in 1962. In this device,

FIGURE 10-3   Picture of the 1-MW solar furnace constructed at Odeillo, France, under the direction of F. Trombe. The north side of the building has been formed into the shape of a large parabolic reflector. (Courtesy of Nicolai Shur.)

the sun's radiation was reflected by a large, flat mirror that tracked the sun. This large mirror consisted of 168 separate plane mirrors of silvered glass, each 50 by 50 cm in size. The solar radiation was reflected from this large mirror onto a second concave reflector 11 m in diameter, made up of 3500 small reflectors, and located 21 m from the first reflector. The parabolic concentrator focused the solar radiation onto a point where 50 kW of power were developed and used both for research and as an industrial smelter. Later, Trombe constructed an even larger, 1-MW (thermal power) solar furnace at Odeillo in southern

France. In this device, the sun's rays are reflected by 63 heliostats, covering an area of 2835 m^2, and located on the side of a hill. The reflected radiation is then directed on a parabolic concentrator designed as an integral part of a nine-story building. This concentrator, shown in Figure 10-3, is 40 m high and 54 m across. The peak flux at the focus of the parabolic concentrator is 16 million W/m^2 or about 16,000 times the standard insolation of 1000 W/m^2. (Let us call this a concentration factor of 16,000 suns). For many years this instrument was used for a wide variety of high-temperature experiments. In 1973, the transformation of the solar furnace into an electric power plant was begun; in November of 1976, solar electricity flowed into France's electric grid. Since then, the facility has been reconverted to one for research at high temperatures.

In Italy, Francia developed an intricate clock-driven field of 271 heliostats and was able to produce steam at a rate equivalent to 150 kW. This concept is not thought to be suitable for large solar-electric plants because of the difficulties associated with connecting many thousands of heliostats into a single, clockwork device. A 400-kW (thermal power) test facility of this type has been constructed at Georgia Technological University. This solar-thermal test facility has 550 heliostats, covering an area of 532 m^2, resulting in a peak solar energy density that is almost 4000 suns.

Experimental central receiver plants are now in operation or under construction in many countries. The first large system was the 1-MW (megawatts of thermal power) plant at Odeillo; a larger 2-MW$_e$ plant has been constructed at Tavgasonne, near Odeillo. In this case molten salt is used both as the heat transfer fluid and as the thermal storage medium. The European Economic Committee built a 1-MW$_e$ plant in Adrano, Sicily and the International Energy Agency constructed a 0.5-MW$_e$ solar tower facility at Almeria, Spain. The Spanish government has also built a 1-MW$_e$ plant at Almeria.

The first major step in the U. S. research and development program to implement commercial solar electric power using the tower approach involved the completion and testing of a 5-MW (thermal) facility located at Albuquerque, New Mexico. A mirror field of 222 heliostats covering 8257 m^2 was utilized. A peak flux of 2.5 million W/m^2 has been obtained, representing a concentration factor of 2500 over the standard value. Figure 10-4 shows a view of the partially completed facility. The tower is 61 m tall and was constructed from 4400 m^3 of concrete. Each heliostat is anchored with nine 10-ton concrete footings, and the entire mirror field covers 40 hectares. The next stage in the federal development program involved the construction of a $140 million, 10-MW$_e$ development plant at Barstow, California. The Barstow facility, described in detail later, is now completed and operating successfully.

FIGURE 10-4 View of the 5-MW (thermal power) solar tower facility at Albuquerque, New Mexico. The mirrors focus sunlight on the top of the 61-m tower. (Courtesy of Sandia Laboratories, Albuquerque, New Mexico.)

B.2. Design Factors

Because the heliostat field comprises almost 50% of the total cost of a solar tower system, good design is crucial. A typical system might consist of a small heliostat mounted on a pedestal and positioned by an individual automatic control system. Each reflecting mirror then images the sun onto a fixed surface. The pedestal supporting the mirrors would have 2 degrees of freedom of motion providing rotation and tilt. Each morning the rotation and tilt positions on the mirrors would be indexed to their proper position, and the automatic control system would then maintain the reflected energy on the receiver located at the top of the tower. For aerospace companies, who are used to designing highly accurate tracking mechanisms for ballistic missiles, the degree of accuracy required here can be easily and inexpensively attained.

In 1975, four contractors in the United States began to build and test first-generation heliostats. Two basic approaches were utilized. Three of the contractors used different types of silver-plated glass mirrors mounted on steel structures. The fourth approach, utilized by Boeing Aircraft Company, involved the use of lightweight plastic reflectors that were mounted inside air-supported domes of clear plastic that protected the mirrors from winds and reduced structural weight. This latter approach suffered from lower optical performance (a 20% energy loss in receiving and transmitting the sunlight through the dome) and from a short life span of the plastic.

Second-generation heliostats are now under test at the Central Receiver Test Facility at Albuquerque, New Mexico. A sketch of a typical design is given in Figure 10-5. These designs, all of the glass and steel variety, are being tested to measure performance and to predict the useful life of the heliostats. Evaluation of test results indicate that the new heliostat designs are all basically sound. Previous design deficiencies have been overcome resulting in better weight efficiency, better pointing accuracy, better beam quality, and longer expected life spans. Expectations are that glass/steel heliostats can be manufactured for less than $125/m^2 (in 1980 dollars) provided a reasonable market is available. It is hoped that heliostat costs in the $100/m^2 range would allow a central receiver system to provide energy at a rate economically competitive with fossil fuel systems.

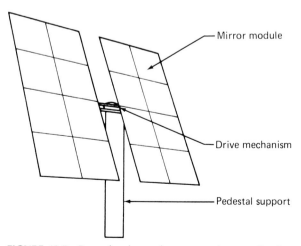

FIGURE 10-5   Example of a modern, second-generation heliostat design. Typical mirrors are about 57 m^2 in area.

The heliostats must be spaced in such a way as to avoid shading and blocking of the reflected radiation. The optimum configuration is obtained with a nonuniform mirror distribution in which about 25% of the land area encompassed by the mirror field is actually covered by mirrors. The density of the reflecting mirrors decreases with distance from the solar tower in order to prevent blocking of the reflected sunlight by adjacent heliostats.

The toughest technical problem to be ovecome may be that of the central receiver. Because of the high-concentration factor, the power density inside the receiver will be quite high and variable. Some designs affect slightly the focus of different groups of heliostats in order to spread out the power profile. The materials used in the receiver must be able to withstand instantaneous changes in energy densities from 0 to 5 MW/m^2. Because of the diurnal variation of the incoming radiation, the system will be subject to frequent temperature cycles. One design concept under study would use a gas, rather than a steam turbine. In this case, the best operating temperature is twice that of a steam turbine. This would rule out the use of metals in the high-temperature face of the receiver; the use of ceramic heat-exchange tubes has been suggested. The use of a dry gas turbine has the strong advantage of reducing the environmental impact as compared to wet systems, which will require the importation of water to the site.

B.3. Solar One

A milestone in the development of solar-thermal technology was reached in 1982. Construction of a central receiver pilot plant capable of generating 10 MW of electricity was completed. This experimental facility is located at the Southern California Edison Cool Water Generating station in Daggett, California, 19.3 km east of Barstow, California. A photograph of the completed facility is shown in Figure 10-6. The photograph was taken in the early evening as the heliostats were being "stowed away" for the night. This facility started providing power to consumers throughout the Southern California electrical distribution grid in the spring of 1982.

The project was launched in January 1977 by the U.S. Department of Energy (DOE). Solar One is a joint venture of Southern California, Edison, and the Los Angeles Department of Water and Power, who invested $21.5 million in the turbine-generating portion of the project, and the U.S. Department of Energy, which funded the remainder at $120 million. This project is the first-of-a-kind for the United States and was constructed to be a pilot operation for judging the technical and economic feasibility of central-receiver generating stations.

Solar One was built despite serious objections by many different groups involving solar enthusiasts who favored low-technology solar aspects such as active hot water systems and have an abiding distrust of high-technology solar projects. Daniel Yergin, the author of Harvard's

FIGURE 10-6 Photograph of the central receiver facility near Barstow, California. The picture was taken in the evening as the heliostat field is being placed in the stow (vertical) position. Mountains are toward the North. (Courtesy of Southern California Edison.)

highly praised "Energy Future" study, described Solar One as "science fiction solar." Fortunately, the Barstow facility is now operating very satisfactorily. The project's success has both technical and administrative people thinking more optimistically about the future of solar electricity.

The plant is designed to produce a maximum of 10.8 MW_e to the power grid for 7.8 hours at June 21 and for four hours at December 21. The heliostats reflect solar energy onto a receiver mounted on a 90.8-m tower. In the receiver, water is boiled and converted to high-pressure steam (560°C) which is then converted to electrical energy by the turbine/generator system. Any steam from the receiver in excess of the energy required (35.7 MW_t) to generate the desired 10-MW_e power for the utility grid is diverted to thermal storage for use when the receiver is receiving energy at a rate below that needed for the rated output.

The collector field, consisting of 1818 heliostats, has a total reflecting area of 72,538 m^2 and is divided into four quadrants. Each heliostat is constructed of 12 slightly concave mirror panels totalling 40 m^2 of mirrored surface. The mirror assembly is mounted on a geared drive unit for azimuth and elevation control. The control system consists of a microprocessor in each heliostat, a heliostat field controller for groups up to 32 heliostats, and a central computer called the heliostat array controller. The annual and diurnal sun position information for pointing each heliostat are stored within this control subsystem. The heliostats can be controlled individually or in groups. The plant operator can control the collector field through either the heliostat array controller or via a master controller arrangement.

Economic evaluations indicate that solar tower facilities can compete very favorably with other energy sources, provided that a modest development effort continues. Over 45% of the total cost of a solar tower plant is for the heliostats. The cost of an individual heliostat for Solar One was $375/m². A reduction in cost to $100/m² can be achieved; this will have a large impact on the eventual energy cost from such plants.

B.4. Energy and Power Evaluation

We end this section with an evaluation of the useful energy and power available from a solar tower facility. The amount of useful energy E_A can be expressed as

$$E_A = I_D A_H \varepsilon_s \varepsilon_c \qquad (10\text{-}1)$$

where

I_D = annual direct solar radiation (MJ/m²)

A_H = A (area of land) × ε_H (fraction of land area actually covered by heliostats) (We use $\varepsilon_H = 0.25$.)

ε_s = efficiency of conversion of the insolation incident on the heliostats into thermal energy. (We use the value of 0.53 quoted in solar tower literature as being representative.)

ε_c = Carnot efficiency (We use 0.39. For an operating temperature of 566°C and a sink temperature of 100°C, the ideal Carnot efficiency would be 0.56. Our value is typical of the reduction in efficiency from the ideal for a real operating plant.)

As an example, assume that the annual direct solar radiation at the site is 7500 MJ/m². In this case, Eq. 10-1 gives

$$E_A \text{ (GWh}_e) = 107.8 \, A \qquad (10\text{-}2)$$

where A is in millions of square meters and E_A is in gigawatt hours (10^{12} Wh). Expressing this result in terms of A in square miles,

$$E_A \text{ (GWh}_e) = 279.2 \, A \qquad (10\text{-}3)$$

Actually, the above results considerably overestimate the amount of useful energy collected. All of the incident solar radiation is not effective in powering the plant. Let us assume that a solar intensity of less than 40 to 50% of the clear day value is below the threshold for operation. Frequency distribution studies then indicate that only 60 to

75% of the total solar radiation intensity lies above the threshold. At best, the above results should be multiplied by 0.75.

To calculate the peak power of the facility, assume that the incident solar intensity has the value of 950 W/m², which roughly is the solar constant attenuated through the earth's atmosphere. In this case,

$$P_M = A_H I_0 \varepsilon_p \varepsilon_e \qquad (10\text{-}4)$$

where

$I_0 = 0.95$ kW/m²,
$\varepsilon_p = 0.64$ for the operating efficiency for the conversion of the incident radiant energy to thermal energy

The other symbols have their usual meaning. ε_p is greater than ε_s (= 0.53) because there is less shading of one reflector by another near solar noon. Also, the geometrical situation is more favorable at this time, as the light will strike the heliostats closer to the normal. With these numbers, we obtain

$$P_M \ (\text{MW}_e) = 59.3 \ A \qquad (10\text{-}5)$$

when A is in square kilometers and

$$P_M \ (\text{MW}_e) = 153.6 \ A \qquad (10\text{-}6)$$

when A is expressed in square miles. This relation is graphically illustrated in Figure 10-7, with the area expressed in square miles.

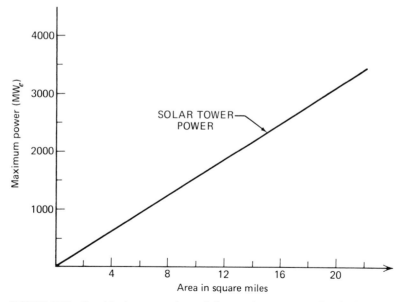

FIGURE 10-7 Graphical presentation of the maximum power level of a solar tower facility as a function of land area covered. In practice only about 25% of the land area will be covered by heliostats.

An interesting comparison between a solar tower facility and a solar-electric plant using solar cells can be made using Eq. 10-1. We will find in the next chapter that a realistic solar cell efficiency is about 10%. But, since the product of $\varepsilon_s \varepsilon_c$ is 21%, the overall efficiency of solar tower facility for converting solar radiation into useful electrical energy is twice that of a solar cell electric-generating station.

C. DISTRIBUTED COLLECTORS

The solar tower concept takes advantage of the high-concentration ratio for the incident radiation, achieving this with a large number of slightly curved reflectors that follow the sun through the use of a two-axis tracking mechanism. Each mirror must be continuously pointed in order to keep the sun's image focused on the boiler. The distinguishing feature of the central-receiver concept is that all of the solar energy is collected at a single point.

An approach that is intermediate between the highly concentrating power tower concept and the nonconcentrating flat-plate collector is to use a series of distributed collectors, each of which has a modest concentration ratio (perhaps 10–100) to collect the solar energy by heating a fluid to some moderately high temperature, usually in the 100 to 500°C range. The basic idea involved here is certainly not new, having been utilized by a number of others in the early part of the twentieth century, such as the Shuman and Boys power plant in Egypt, which developed 50 hp for an irrigation project. However, these early engines all suffered from very low overall efficiencies of the order of 3% or less.

Modern technology offers the possibility of a number of improvements over earlier solar designs. Moderate concentration ratios can be obtained through the use of well-constructed parabolic reflectors. These can either track the sun and focus the sun's rays on a stationary receiver or, alternatively, a stationary reflector can be combined with a movable receiver. Losses in the receiver can be minimized through the use of modern techniques such as an evacuated inner element combined with a selective surface. This would eliminate convective losses and greatly reduce radiative emission in the infrared from the hot fluid. In either case, a single-axis tracking mechanism is utilized.

Distributed collectors operate by collecting sunlight and converting it to heat at each individual collector module. Some common distributed collector concepts are illustrated in Figure 10-8. Each module has its own heat collection pipe (equivalent to the central receiver in the tower approach). Collection areas are typically a few square meters, so that many thousands of these modules would be required to provide the power output equivalent to a 100-MW solar tower plant. The large

FIGURE 10-8 Solar Systems Test Facility at Sandia Laboratories showing the solar collector fields designed and fabricated by Suntec Systems, Inc. (right foreground) and by General Atomic Company (right background). The Suntec collector arrays have mirrored slats that track the sun and concentrate solar radiation onto a fixed received assembly. General Atomic's collectors have a stationary precision-cast concrete trough with silvered glass facets that focus the sun's rays on a sun-tracking receiver tube. The sun-tracking collectors on the left were designed and fabricated by Sandia. (Courtesy of Sandia Laboratories, Albuquerque, New Mexico.)

amount of interconnected plumbing required appears to rule out the distributed collector approach for very large installations. For small electrical power plants and for systems providing both power and heat, the distributed collector concept is becoming increasingly attractive.

Most of the distributed collector designs operate on the direct component of the incoming solar radiation and must track the sun across the sky. Since the distributed collector modules have only a single-axis motion, accurate aiming of the concentrator at the sun is important. Good sun-tracking subsystems should be sufficiently accurate to keep the pointing error equal to or less than 0.10°. This type of precision will keep losses due to misalignment below 2%. The automatic tracking system should include the capability to quickly detect a loss of insolation. Also, it must be able to reposition the collector, at the end of the day, to the proper orientation for collecting solar energy the next morning.

Test data for one type of distributed collector system are shown in Figure 10-9. The data show that good collector efficiencies can be obtained, provided that the thermal losses at higher operating efficien-

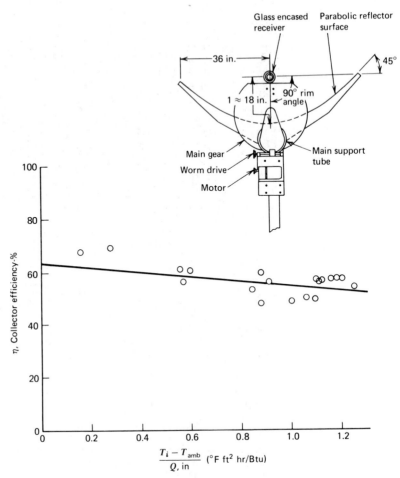

FIGURE 10-9 Summary of results obtained for a typical distributed collector using a parabolic concentrator and a selective surface on the receiver. Insert shows details of the design.

cies are minimized. The theoretical curve was calculated assuming a selective surface with an infrared emissivity of 0.10. These data emphasize the importance of low emissivity in reducing the thermal losses due to infrared reradiation.

A system of distributed collectors lends itself readily to a wide variety of low-power, on-site applications. This type of system can pump water for irrigation, produce steam for industrial processes, generate electricity in small- and medium-sized installations, and also supply heat for residential use. These systems may prove most economical when several purposes are served in a complementary manner. One particularly attractive application is to provide process heat for the food, textile, and chemical industries.

Most of the equipment needed for such systems is now available. Development of concentrating collectors that can raise fluid temperatures above the boiling point of water has progressed particularly rapidly, at a cost generally comparable to that for simpler, flat-plate collectors. The design aspect most in need of further developmental work is that for small heat engines to convert solar heat into shaft power to drive a generator, air-conditioning compressor, or water pump. The cost of one-axis, tracking collectors is of the order of $50 to $200 per square meter as compared to the present cost of $500 to $1000 per square meter for the more complex, two-axis tracking collector systems. The custom-made heat engines used today cost about $1000 per kilowatt, but mass production techniques should reduce this to a figure below $200.

Distributed collector approaches lend themselves most readily to small power system applications. A small power system is commonly defined as one where peak power is less than 5 MW_e. Naturally, small solar tower plants could also be used for this purpose. A schematic diagram of a small solar power system is shown in Figure 10-10. This particular approach utilizes parabolic dish systems which require two-axis tracking. In the scheme shown in Figure 10-10, cavity receivers are mounted at each dish and the hot steam is carried away to a central turbine and storage system. A possible variation on this approach is to mount a heat engine and generator at the focus of each reflector. This approach requires many small heat-engine units and they must be cost effective as compared to one large central unit.

Many different single-axis tracking collector systems are also possible. The parabolic trough system would normally use steam for the

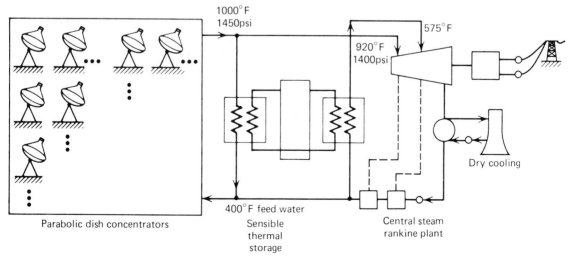

FIGURE 10-10 Schematic diagram of a parabolic dish distributed collector system.

FIGURE 10-11 The solar-powered irrigation project at Willard, New Mexico. The parabolic solar collectors are shown in the center of the photograph, and the thermal storage tank and heat engine are on the left side of the collectors. The storage pond is in the foreground. The system is used to irrigate 40 hectares of different crops. (Courtesy of Sandia Laboratories, Albuquerque, New Mexico.)

energy transport to a central turbine-generator unit, although any of these one-axis tracking systems could use small turbine generators associated with either individual collectors or with small groups of collectors. It is not necessary to transport the energy to a central conversion point, although right now the central collection approach appears to be the most practical alternative for steam conversion systems. If the conversion to electrical power is done at the individual collectors, there would no longer be the option of using thermal storage with each unit. Instead, it would be necessary to install an electrical or electrical intermediary storage system, such as pumped hydroelectric power. When collecting heat to a central point from a parabolic trough, the best form of storage would normally be a thermal storage system. The final conversion to electricity would probably be done using a conventional steam turbine.

A typical practical application of small solar systems is for irrigation. It has been estimated that there are over 300,000 irrigation pumps in use in the western United States, most requiring 50 kW or more. One of the first intermediate-temperature projects to get under way was for pumping water. Northwestern Mutual Life, which has a large farm at Gila Bend near Phoenix, Arizona, undertook a 38-kW (50 hp) irrigation project in collaboration with Battelle Memorial Institute. The project was initiated in August 1975, and the system began pumping water 18 months later. At the peak of insolation in June, the system pumps 10.6 million gal of water per day. This system has 550 m^2 of collector surface shaped in the form of parabolic troughs. A larger plant with 2140 m^2 of collector provides electricity for pumping water from three 91-m-deep wells at Coolidge, Arizona to irrigate cotton crops. A photograph of a typical installation is shown in Figure 10-11.

A recent example of the distributed collector approach is the Small Solar Power Systems (SSPS) project in Almeria, Spain. This project is supported and funded by nine different countries including the United States. This SSPS plant, which has a rated output of 500 KW$_e$, took advantage of the pilot-system experience gained with the irrigation projects. The design incorporates two collector fields of approximately equal size. One field comprises 10 loops of 60 collectors manufactured by the U. S. Company, Acuvex. The other field consists of 14 loops of 6 collector modules developed by a West German manufacturer. The stated objective of the project is to determine the feasibility of solar electricity generation in remote areas whose geographical situation renders conventional electricity both difficult and costly.

D. OCEAN THERMAL-
ENERGY CONVERSION

In marked contrast with the above proposals are schemes that would take advantage of the small but important difference of temperature between the surface water of tropical oceans and the cooler water at a depth of 300 to 600 m or more below the surface. The ocean's thermal resource is enormous because of the distribution of water in the tropical regions. About 50% of the solar energy intercepted by the earth falls between the Tropic of Cancer and the Tropic of Capricorn. In this area, 90% of the earth's surface is covered by sea, so that 45% of all the earth's incoming solar energy is absorbed by the tropical oceans. This vast reservoir of water contains enough to supply many times the world's energy needs. The surface layer, which is a few hundred feet thick, absorbs the energy from the sun and stays at about 25°C. Cold water from the poles slides into the depths of the oceans and slowly moves toward the equator, providing a cool reservoir at a temperature of 5°C and a depth of the order of 600 m. Generally, the cold and hot layers do not mix, thereby providing a warm heat source and a cool sink for running a thermal engine.

A large fraction of the cost for most solar electric schemes is involved in the construction of the apparatus to collect the solar energy. In addition, large and expensive arrangements for thermal storage will be required. Power plants that take advantage of the ocean thermal gradients are projected to be very cost competitive with more conventional power plants because the sea acts as the medium for both the collection of solar energy and the storage of energy. The major fraction of the cost arises from the cost of construction of the thermal engine and from the need to transmit the resulting electrical energy from the plant at sea to the shore.

The idea of using ocean thermal gradients to generate electrical power is not a new one, having been first suggested by the French physicist D'Arsonval in 1881. In 1926 the French engineer Georges Claude designed a 60-kW turbine, which worked using a difference of only 20°C in the working fluid. In 1930 he attempted to build a 40-kW power plant in the tropical sea waters off Cuba. After numerous difficulties, the plant did work in a fashion, producing 22 kW from a test engine operating on a 14° temperature gradient, although whether this was a net gain of power for his plant is not clear, since energy was required to pump water up from the depths. Claude used water as the working fluid.

Because of the small, 20°C temperature differential provided by the ocean, the maximum possible efficiency would be about 6%. In fact the operating efficiency is probably closer to 3% because of various losses associated with practical heat engines.

The low-conversion efficiency of the solar sea power plants leads to a serious problem. Since the thermal efficiency of the system is less than one-tenth that of a conventional fossil fuel plant, the heat exchanger within the boiler must transfer more than 10 times as much heat in order to obtain an equivalent power output. This implies that the boiler must be 10 times as large and 10 times as expensive. However, two other factors must be considered that tend to overcome this cost increase. The boiler tubes in a conventional power plant operate under the doubly adverse conditions of high pressure and temperature, requiring thick walls made of expensive materials. By contrast, the sea power plant boiler operates at low temperatures and pressure, so that much thinner boiler tubes can be used. In a submerged system, the sea water's hydrostatic pressure would largely compensate for the vapor pressure of the working fluid. With ammonia, the vapor pressure of the boiler would be balanced by the ocean's hydrostatic pressure at a depth of 60 m. In other words, by being able to install the boiler and condenser underwater at convenient depths where water pressure on the outside can equalize internal pressures, construction can be relatively "flimsy" with thin-walled tubes throughout.

The operation of such a sea power plant is illustrated using the schematic shown in Figure 10-12. Warm water is pumped into the boiler to boil the working fluid, shown as ammonia in this case. The ammonia gas under "high pressure" is fed into a turbine generator and is discharged at "low pressure" into the condenser, which also receives cold water from the deep ocean. Ammonia liquid at low pressure from the condenser is then pressurized and pumped to the boiler where the cycle is then repeated.

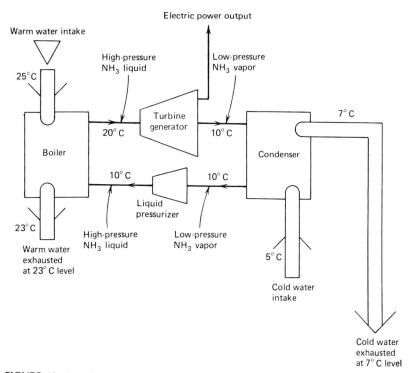

FIGURE 10-12 Schematic arrangement of a solar power plant operating on the small ocean temperature gradients existing in tropical waters.

Many problems remain to be solved before practical power is achieved. These include fouling of the heat exchangers by marine organisms, mooring versus dynamic positioning, which is the best working fluid, corrosion of the metal parts, and so on. Those working in this area are quite enthusiastic about the possibilities of this approach because it is a field that does not appear to require any major technical developments. Further justification for enthusiasm was given in 1979 by the successful demonstration of the ocean thermal approach in an experiment near Hawaii. In this pilot experiment the ocean temperature differences drove a 50-kW generator.

E. SOLAR PONDS

In 1948 Rudolph Block suggested that an effective solar collector could be created by avoiding convection in a stratified salt solution. This phenomenon had been observed as early as 1900 in several Hungarian lakes. This discovery was made by A. V. Kalecsinsky, who was puzzled by the unusual temperatures in Lake Melve. Water with a temperature of 70°C (150°F) was found at the bottom of the lake. Kalecsinsky noted that solar radiation absorbed in the body of these lakes resulted in an increase in temperature with depth. Because of a natural salt-concentration gradient in the lakes, lower regions remained denser even when warmer. Convection is prevented because the increase in density offsets the effect of thermal expansion due to local heat absorption. When no convection can occur, water can lose heat only by conduction, and this heat-loss mechanism is slow. One meter of nonconvective water is about as good an insulator as 6 cm of styrofoam. Except right at the surface, radiation losses are zero, since water is opaque to low-frequency, infrared radiation. In natural solar ponds the salt-concentration gradient is maintained by means of salt deposits at the bottom of the lakes that provide a close to saturated solution there and by fresh water streams that flow across the top of the lakes.

Under suitable conditions, it appears likely that even higher temperature rises might be attained. Thus, the possibility of utilizing the solar energy collected in the form of heat at the bottom of such a pond becomes more practical. In addition to its relative simplicity, the solar pond has the attractive feature of having a long-term storage capacity built into the solar collector, in contrast with the few days of heat storage that is usually available with water tanks and rockbeds in more conventional systems. This suggests its application to space heating, particularly to housing complexes where a large solar pond might provide sufficient heat for several buildings.

Experimental studies on solar ponds were carried out in the early

1960s in Israel by Tabor and collaborators. A 25-m solar pond with a blackened bottom was constructed on the shores of the Dead Sea. Using an artificially created salt-concentration gradient, they were able to obtain a temperature of over 90°C at a depth of 0.8 m. The goal of this study was to achieve economical electrical power. In his original paper, Tabor felt that power generated using the solar pond concept might prove competitive with commercial fuels.

The principle of a solar pond is illustrated in Figure 10-13, where a regular pond of water that is homogeneous throughout is compared

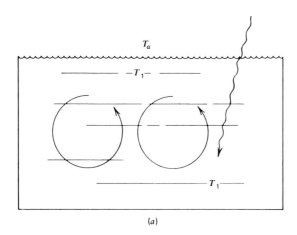

(a)

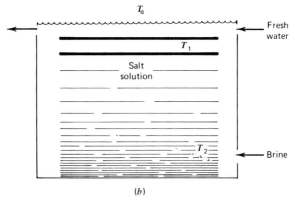

(b)

FIGURE 10-13 A nonconvecting solar pond is contrasted with an ordinary water pool. The temperature of the water in the homogeneous pond in (a) is essentially constant throughout, resulting from the circulation of water. In the solar pond shown in (b), the stratified salt solution stops all convection and permits a temperature gradient to be attained with $T_2 > T_1$.

with a solar pond in which a salt-concentration gradient is maintained with the lighter, fresh water on top and the heavier, salt water at the bottom. Both ponds are sufficiently transparent that a large fraction of the solar energy penetrates to the bottom and is absorbed there. As the water at the bottom of the homogeneous pond warms up, it expands and rises to the surface where it cools off again. In the solar pond the thermal expansion is insufficient to disturb the stability provided by the salt-concentration gradient and convection is suppressed. Since the heat loss upward by conduction is very slow, the temperature at the bottom rises and a permanent temperature gradient is established. Useful heat can then be extracted from the bottom of the pond, although care must be taken not to upset the stability of the solar pond.

Because the incident solar radiation covers a wide spectrum of wavelengths, only a portion of it will penetrate to the bottom of the pond. About 24% of the insolation lies in the relatively short-wavelength region of 200 to 600 nm. At a depth of 1 m, only one-eighth of this has been absorbed by the intervening solution, so that most of this 24% will be absorbed by the bottom layers or by the blackened bottom. About 35% of the incident solar radiation lies in the wavelength interval of 600 to 900 nm; of this one-third penetrates a depth of 1 m. The remaining 40% of the incident solar radiation is absorbed in the first 10 cm of depth because of the opaqueness of long-wavelength radiation in water. Even without detailed calculations it is evident that the optimum depth of a solar pond lies between 1 and 3 m.

The warming of a solar pond is shown, in Figure 10-14, to be a function of the time after filling. In steady-state conditions the pond follows the pattern of the incident solar energy with a lag in phase of about two to three months. This conclusion was reached in both instances as a result of both theoretical analysis and some limited experimental observations.

The problem of keeping the pond clear is a serious one, although probably simpler than trying to keep equally vast areas of glass or mirror clean. If the dirt consists of floating particles, these particles can probably be removed by the surface-washing process. Heavier particles that sink to the bottom will cause no problem unless they reduce the effectiveness of the black absorber on the bottom. Particles that remain suspended are more serious, since the effective transmissivity of the pond will be lowered. Biological growth can be suppressed by chemical means.

The rate of molecular salt diffusion is remarkably small. It is claimed that no action needed to be taken to correct for diffusion in the small experimental ponds. To maintain the gradient it is only necessary to inject a concentrated brine solution into the bottom of a pond from time to time. The top of the pond is washed with a weak solution with some overflow. An equilibrium concentration will be created at the top so

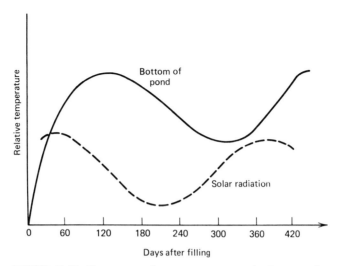

FIGURE 10-14 The expected temperature near the bottom of a solar pond is shown as a function of the time after filling. Also illustrated is the phase relationship between the pond temperature and the insolation. Roughly speaking, there is a lag of about 60 to 90 days between the two quantities.

that the salt carried away in the overflow equals the salt diffusing upward through the pond.

Whether the ground below the pond can provide additional storage of heat depends on the level and flow rate of groundwater and on the size, since ground storage is feasible only if the pond is large compared to the distance the water moves per year. If the ground leakage is fast, the bottom and sides of the pond should be insulated. Otherwise ground storage is an attractive feature, since it offers heat storage equivalent to nearly 1 m of water at no extra cost.

For a small pond, the heat losses from the boundaries are excessive. As the pond is made larger the effect of the sides decreases, and the heat loss from the bottom vanishes after the pond has heated (provided that groundwater flow rates are not excessive). Weinberger has shown that a small pond gives a lower ultimate temperature rise than an infinite pond by a factor of approximately

$$\frac{\Delta T}{\Delta T_{max}} = 1 - \frac{2h}{D} \tag{10-11}$$

where h is the depth and D is the diameter. This means that, for a pond that is 1 m deep, the temperature rise will be within 90% of that for an infinite pond if D is larger than 20 m.

Extraction of useful heat from a solar pond is a problem whose solution clearly awaits future experimental work. The simple solution of an array of pipes through which a heat-exchange fluid such as water is piped may not be practical for the type of pond outlined in Figure 10-13. For useful heat exchange to take place, it is necessary that some convection outside the pipes occur. Nevertheless, the danger of convection is that it might destroy the stability of the pond. Another possibility is to extract liquid from the bottom of the pond, pass it over an external heat exchanger, and then return the liquid to the pond. Because of the density gradient it is known that a layer can be removed without causing mixing. The need to return the extracted fluid, which will now be well mixed, requires a homogeneous convecting layer at the bottom. A variation on this possibility is to construct the solar pond with a glass or plastic partition between the upper insulating layer, with its necessary salt gradient, and the bottom homogeneous layer, which is convecting.

The pioneering work in Israel was stopped in the early part of the 1960–1970 period because of the low cost of oil and was not resumed until after the oil crisis of 1973. This second effort was lead by Tabor and by Ormat Turbines, Ltd. one of the largest manufacturers of Rankine cycle turbines for electric power generation. Ormat was formed by L. Bronichi, who spearheaded an effort to develop turbines for solar-powered pumping of water in arid lands. This generator was small, but had two unique features: It needed only 90°C water to produce electricity and was very reliable. The only moving part was the turbine. The bearing were lubricated with the same organic fluid that powered the turbine.

Ormat, in collaboration with the Israeli government, produced several working solar ponds. The largest, a 7000-m^2 pond, began operation at Ein Bokek in December 1979. It can produce 35 kW continuously or 150 kW peak power. A larger 5-MW_e facility should be in operation sometime late in 1983. This is the first of a series of salt-gradient solar ponds that are envisaged for the Dead Sea. The eventual goal is to produce 2000 MW_e by the end of the century.

F. ENERGY STORAGE

A suitable system of energy storage is perhaps more crucial to economic solar energy development than for conventional approaches. Since the sun only shines for a portion of each day, a means to provide short-term storage of energy is an essential ingredient in any solar collection system. The common occurrence of inclement weather, particularly in northern climates, demands some type of longer-term storage. In this section several approaches to energy storage are reviewed.

The most common approach to energy storage involves the storage of thermal energy as sensible heat, that is, heat is stored by raising the temperature of a given mass of material. We have already discussed the use of water and rock for heat storage at low temperatures in Chapter 9, where it was noted that on a weight basis, water could store about five times as much heat as rock for a given ΔT. Because rock, or concrete, is considerably denser than water, there is only a factor of about 2 difference between the two types of storage on a volume basis. As pointed out earlier, rock storage has the advantage of allowing more stratification of the heat, thus permitting more efficient extraction of heat from the storage medium. For high-temperature thermal storage, water becomes much less attractive because of its low boiling point. Since the ability of iron oxide to store sensible heat is about 90% that of water on a volume basis, this may be a more attractive approach for high-temperature thermal storage of heat. The disadvantages of sensible heat storage systems for large-scale systems are that the storage medium takes up a lot of space, and the heat cannot be stored effectively for long periods of time.

A good example of a large sensible heat storage system is that employed at the Barstow central receiver facility. The main storage is a 3581-m^3 (946,000 gal) tank filled with sand and about 908 m^3 (240,000 gal) of oil. Hot steam from the receiver is routed through heat-exchange coils located in a tank containing some of the thermal storage oil. The heated oil is then pumped back into the main storage tank where thermal energy is transferred to the rock and sand. When fully charged the temperature of the oil, sand, rock mixtures is about 300°C. The rated electrical capacity of the plant operating only on stored heat is 28 MWh.

Another method of thermal storage is to make use of the latent heat associated with a phase change in a material, such as the transition from a solid to liquid phase. For example, it takes 540 cal to change 1 g of water from a liquid to hot steam. Unfortunately, gaseous mixtures are usually difficult to handle so that only materials undergoing a phase change from a solid to a liquid have been given practical consideration. In general, phase changes from a solid to a liquid require less energy than for a change from a liquid to a gas. The heat of fusion of sodium chloride, which undergoes a phase change near 800°C, is 123 cal/g. The advantage of this type of thermal storage over sensible heat storage is that a great deal more heat can be stored in the same volume of material. Numerous difficulties with this approach remain to be overcome. The materials are usually expensive and corrosive. Some materials have short lifetimes due to decomposition when thermally cycled. Problems associated with supercooling have plagued the use of these materials in low-temperature applications, as the amount of heat that can be stored decreases after the heat-of-fusion salt has been thermally cycled a few times. For higher-termperature applications, eutectic salts consisting of a mixture of materials (e.g., sodium nitrate and sodium chloride) may prove useful.

Storage of mechanical energy by pumping water to a higher elevation is a technique widely used by the power industry to store off-peak power in the eastern part of the United States. The stored energy is then released at peak demand hours. The energy required to pump the water uphill is typically recoverd with an efficiency of about 70%. However, the number of sites available for this approach is limited. On a large scale this would probably require moving too much water to be practical. Perhaps a better method of mechanical storage involves the use of rotating flywheels. The stored energy in this approach can be increased either by increasing the mass of the flywheel or by increasing its rotational velocity. The principle of the flywheel has been known for thousands of years, but its use has been limited by the fear of flywheel failure at high-rotational speeds, resulting in large chunks of material rocketing off and causing great damage. The real future of flywheels will depend on the development of flywheels made up of many long, thin fibers. These flywheels would be able to withstand very high rotational velocities and would have the nice property of disintegrating into dust in the case of a flywheel failure.

Storage of energy by chemical batteries has always appeared as an attractive possibility. To be useful in connection with solar energy storage, the battery needs to have a large storage capacity and should have a long lifetime under operating conditions involving a large number of charge-discharge cycles. Presently the best batteries can stand less than 1000 charge-discharge cycles, whereas a lifetime characteristic of 10,000 cycles is probably needed. It is presently believed that a battery that could deliver a specific energy of 220 Wh/kg, having a specific power of 55 W/kg and a storage efficiency of 70%, would be attractive for large-scale energy storage. One attractive possibility which is undergoing research now is the sodium-sulfur battery. For efficient operation it must be operated at about 300 to 350°C. The unique feature of these cells is the electrolyte, which is a solid, ceramic material called beta alumina.

We conclude by noting that the possibility of operating under a "hydrogen economy" has received a lot of attention in the past few years. The hydrogen (and oxygen) would be generated by electrolysis when an electric current is passed through water. The hydrogen could be stored or transported to appropriate sites by pipelines and recombined with oxygen as with fuel cells; an efficiency of 50 to 60% for recombining hydrogen and oxygen to produce electricity is possible.

REFERENCES

1. Kenneth W. Ford, *Basic Physics*, (Waltham, Mass.: Gin and Blaisdell Publishing, 1968).

   An elementary discussion of thermodynamics can be found in numerous introductory texts on science and engineering. However, the presentation by Ford is extremely thorough and penetrating.

Even someone with an advanced knowledge of the subject can benefit from the original and thoughtful discussion given in this textbook.

2. *Sunworld*, Vol. 5, Issues 3 and 4. (New York: Pergamon Press, 1981).

Sunworld is an official publication of the International Solar Energy Society. These two issues are entirely devoted to the topic of Solar Thermal Power.

3. A. F. Hildebrandt and L. L. Vant-Hull, "Power with Heliostats," *Science*, 197, 1139 (September 1977).

A good technical description of the solar tower concept as envisaged by the University of Houston group.

4. H. Tabor, "Solar Ponds—Large-Area Solar Collectors for Power Production," *Solar Energy*, 7, 189 (1963).

Contains a detailed discussion of the problems involved and the economic aspects. Describes a number of experimental results.

5. Frank Keith and Jan F. Kreider, *Principles of Solar Engineering*, (New York: McGraw-Hill, 1978).

An excellent presentation of the engineering aspects of solar-thermal devices is contained in Chapter 4 of this text.

QUESTIONS

1. In the distributed-collector solar-electric scheme the magnification (concentration) factor will be of the order of:

    (a) 1   (c) 50
    (b) 5   (d) 1000

2. In the power-tower, solar-electric approach the magnification (concentration) factor will be of the order of:

    (a) 1   (c) 50
    (b) 5   (d) 500

3. Since the solar tower design is more expensive, what possible advantage does it have over the distributed collector (using many trough-type collectors) arrangment? Explain.

4. Is the solar tower approach an example of a single-axis or a double-axis tracking system?

5. For the same area of collectors, the solar tower electric plant would be more expensive than a system using a large number of trough-type collectors. Why?

6. What is the most expensive component of a solar tower electric plant?

7. Is the cylindrical-trough solar collector an example of a single-axis or a double-axis tracking system?

8. Why will the area of the heliostats for a solar tower electric plant amount to only about one-quarter of the total land area covered?

9. What is meant by a heliostat?

10. Name at least three ways of generating electricity from solar energy.

11. An ocean thermal energy conversion plant is moored offshore. List ways to transport the energy ashore.

12. The heat capacity of water is 1.0 cal/g-°C, while that for rock is about 0.2 cal/g-°C. If the density of rock is 2.5 times that of water, what is the relation between the capacity of water to hold heat per unit volume as compared to that for rock?

13. List some advantages and disadvantages of enclosing the plastic tracking heliostats for the power tower scheme in a plastic bubble as proposed by Boeing Aerospace Corp.

14. Explain how a flywheel design, where the wheel is constructed out of fibers pointing in the radial direction, could result in greater allowable rotational velocities than with a conventional design where the wheel is constructed out of a solid piece of metal.

15. What is meant by "high-quality energy"?

16. Some proponents of active and passive solar applications argue that high-technology solar approaches are not needed. If all of our energy needs were supplied through solar energy devices, explain why low-technology solar applications would not be sufficient.

17. Assume that 24% of the U.S. energy requirement is for space and water heating uses (for buildings). Assume the breakdown between hot water and space heating to be 1:2, respectively. Estimate the maximum percentage of the energy usage in the United States that could be met by low-technology solar applications.

11 photovoltaic conversion

In the last chapter the generation of electricity through the intermediary of some type of thermal process was discussed. These thermal approaches appear promising, but they do require large receivers and turbines to generate electricity. In addition, the efficiency of thermal devices is subject to the constraint set by the second law of thermodynamics. A totally different approach involves the direct conversion of sunlight to electricity by solar cells (photovoltaic devices).

Solar cells consume none of our precious energy resources, are efficient and long-lasting, make no contribution to noise pollution, have the reliability inherent in a system with no moving parts, do not represent a health hazard, and contribute no harmful waste products. With all this it should be no surprise if solar cells take care of an ever-increasing portion of the world's energy needs in the future. Already, solar-cell arrays have proven to be economic, reliable, and long-lived energy sources for remote installations such as microwave repeater stations.

Photovoltaic solar cells were used to convert sunlight directly into electricity as early as 1954. However, the prohibitively high cost, roughly ten thousand dollars per peak watt of power, ruled out their use for all but the most exotic applications. Early technical advances in the photovoltaic conversion field resulted primarily from experience gained from the use of solar cells to power satellites and other craft sent aloft by the National Aeronautics and Space Administration (NASA).

For the space effort, reliability and efficiency were the major objectives. As a result, the array cost in 1965 was still over one thousand dollars per peak watt. For this application, the solar cell has proven practical and reliable. Some solar cell systems have lasted over 10 years in space and over 15 years on the ground without serious deterioration.

Solar cells have many advantages. Some of these are as follows:

1. There are no moving parts to wear out.

2. Most solar cells are made from silicon, one of the most plentiful materials on earth.

3. The overall conversion efficiency to electricity is in the 10 to 20% range.

4. The semiconductor industry is a well-developed, high-technology industry. It is probable that solar cells can readily be adapted to mass-production techniques.

5. As the technology improves, the expected lifetimes may become very long. This offers an exciting economic advantage for the long term.

6. Solar cells have a high power-to-weight ratio.

7. Solar cells can be used at the site in many cases, thus helping cut down on the costs of power distribution systems.

The major obstacle to the immediate widespread use of solar cells is their enormous cost. In 1973 the cost per peak watt of power from solar cells was still approximately $300. (A peak watt is defined as follows: the solar cells are required to produce 1 W when the insolation level is 1000 W/m^2. If a solar cell has a conversion efficiency of 12%, then 1 m^2 of cell area corresponds to 120 peak watts.) As with so many other aspects of federal policy, the 1973 Arab oil embargo caused the government to change its outlook on photovoltaic energy conversion. As a result of increased federal expenditures for photovoltaic development (federal expenditures in this area for 1977 reached nearly $60 million), the price per peak watt of power from solar cells began to drop dramatically. Solar cell array costs of $15 per peak watt were achieved in 1977. Although not yet economically competitive with other forms of electric generation, this represents a dramatic reduction in cost from the early days of the photovoltaic effort. The progress in reducing the cost of solar cells is shown in Figure 11-1. By 1982 the cost was about $10 per peak watt.

It must be emphasized that the striking reduction in the cost of solar cells illustrated in Figure 11-1 over the past 15 years cannot continue. We are now at the point where further cost reductions will come only after considerable effort. At $10 per peak watt, solar cell power is too

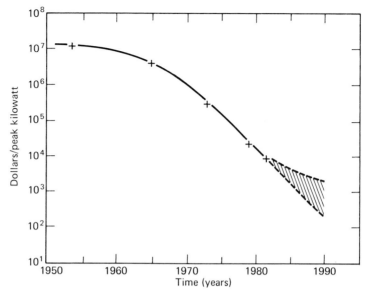

FIGURE 11-1 The cost of solar cells has dropped dramatically from the time of 1953 when they were used only for the space effort. The shaded area from 1982 on illustrates the likely maximum and minimum ranges in future cost reductions.

expensive to compete with other energy sources. A price of $1 per peak watt would make solar cell power systems the preferred alternative to conventional electricity in most cases.

EXAMPLE 1

Let us crudely estimate the present cost of energy using solar cells. Assume that the cost is $10 per peak watt of electrical power and that 1500 kWh of energy per month is desired in a climate that has an average of 120 hr/month at an insolation of 1000 W/m². The cost is found to be

$$\frac{1500 \text{ kWh}}{120 \text{ hr}} = 12.5 \text{ kW (peak)}$$

$$\text{Cost} = 12,500 \text{ W} \times \$10/\text{W} = \$125,000$$

Assume that the solar cell array lasts for 20 years. How much will the energy cost be (neglecting interest on the capital investments)?

$$\frac{\$125,000}{20 \times 12 \times 1500} = \$0.35/\text{kWh}$$

This is cost of 35¢/kWh, and the cost of capital has been neglected, as have been the cost of supporting systems, maintenance, and so on.

These latter items could easily double the total price. It is clear that solar cell costs must be reduced by at least a factor of 5 to 10 before widespread use will occur.

Enthusiasts are hopeful that photovoltaic power will be competitive with other forms within the next 10 to 20 years. Specific plans are to achieve a cost of about 50 cents per peak watt by 1986. How can this large reduction in the cost of solar cells be achieved? One way would be to accomplish a major breakthrough in the research and development program. As we will see later, work is proceeding in the promising areas of thin-film, amorphous-silicon solar cells. Also, significant reductions in the manufacturing costs of the common single-crystal solar cell can probably be made. For example, a reduction in cost to a factor of about 3 is possible through improvements permitting a lower grade of silicon to be used for the starting material. It has been estimated that another significant factor could be obtained through improvements in the methods of growing single-crystal silicon. Finally, automatic fabrication of solar cells might lead to another significant reduction in cost. A major difficulty of the solar cell industry of today is that it is small and production involves a lot of handwork.

Our objective is to learn the basic principles that govern the behavior of solar cells and to study the progress that is being made to develop efficient, low-cost devices. This discussion will be presented in the following manner. First the basic principles underlying the operation of all photovoltaic devices will be examined. This will be followed by a discussion of some of the most promising approaches such as single-crystal silicon solar cells, amorphous-silicon devices, concentrating arrays, and special applications.

A.   BASIC PRINCIPLES

The photovoltaic method does not convert energy first to heat and then to electricity, but instead solar radiation is converted directly to electrical energy. As a first step toward understanding this process it is necessary to consider the energy content of electromagnetic waves such as light. In 1905, physicist Max Planck showed that light must be thought of, not only as a wave, but also as discrete bundles of energy called photons. The energy of each photon varies according to the frequency of the wave and is given as

$$E = hf \qquad (11\text{-}1)$$

where h is a constant (called Planck's constant). Since f varies inversely as the wavelength, the energy content of a photon corresponding to

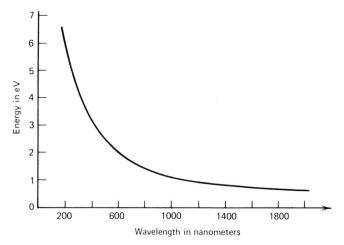

FIGURE 11-2 Relationship between the energy of solar radiation in electron volts and its wavelength.

waves in the blue wavelength region is greater than that for waves in the red portion of the spectrum. It is convenient to describe the energy in terms of a unit called the electron volt (eV). One electron volt is defined as the energy gained by an electron when it is accelerated by a voltage difference of 1 V. In these units the energy of a photon at 300 nanometers (nm) (in the ultraviolet) is 4.15 eV, that of a photon of 600 nm (red) is 2 eV, and that of a photon corresponding to waves of 1200 nm (infrared) is 1 eV. This relationship between energy in electron volts and wavelength is illustrated graphically in Figure 11-2.

For a practical solar cell it is necessary that the photons of the incident radiation be absorbed in some material and liberate electrons in such a way that current will flow in an external circuit. This is most usefully accomplished with semiconductors made of silicon. Silicon has a chemical valence of 4, which means that of the 14 electrons that are normally found in each silicon atom, four are available to interact with other atoms. In a silicon crystal each silicon atom shares its four valence electrons with four neighboring silicon atoms and, these in turn, share the neighboring electrons in the manner illustrated in Figure 11-3. If the structure of the silicon crystal were perfect, each electron would be held in position by the electrical forces between the electron and the two neighboring atoms who share it. Since no electrons would be free to move, a good insulator would result. This is to be contrasted with the case of metals where many electrons are free to move about so that an electric current results when a voltage is applied.

It is instructive to think of this situation in a different manner; one

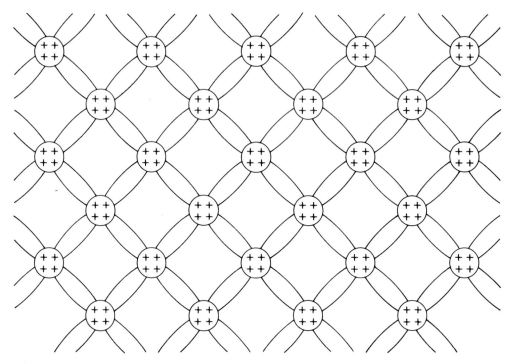

FIGURE 11-3 Covalent bonding gives rise to the regular structure of the silicon crystal. Each atom is firmly held in the crystal lattice by sharing two electrons with each of four equidistant neighbors.

that can be termed a quantum description of the crystal. Figure 11-4 shows a convenient way of representing this situation. This is a band-structure diagram and is interpreted as follows. All electrons that take part in the valence-bonding process in the lattice are said to be in the valence band. If sufficient energy is provided to get them out of the valence band—that is, to break the bonding—they are then free to move and are said to be in the conduction band. The region in between is called the forbidden region or band gap. The height of this gap represents the amount of work necessary to remove an electron from the valence band and put it into the conduction band. The vertical axis represents electron energy, and the horizontal axis indicates position in the crystal structure (in one dimension). Electrons in the conduction band may also have kinetic energy because they are moving. Electrons with different kinetic energies are shown at different distances from the band edge.

The promotion of an electron to the conduction band leaves behind an atomic site that lacks an electron. This situation is commonly called a "hole." When a hole has been created in some way, a neighboring electron can move over to the hole site, leaving behind a new hole. The

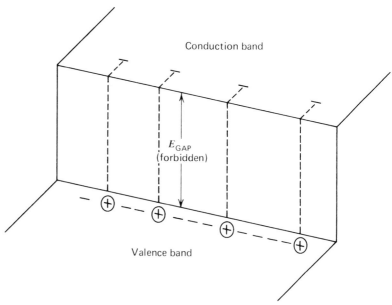

FIGURE 11-4   Schematic representation of the concept of band structure in solids. The significance of the energy gap is discussed in the text.

movement of a bound electron from a neighboring site into this hole constitutes a net motion for the hole. The flow of electric current in the cyrstal is made up of the motions of both electrons and holes. The hole motion, of course, will be in the opposite direction to that of the electrons.

How do electrons get into the conduction band? In a normal crystal at any positive temperature there is a certain amount of random motion of the lattice atoms. Because of this thermal motion, occasionally an electron will be excited from the valence to the conduction band, leaving behind a hole. Clearly, the smaller the gap energy, the more electrons there will be in the conduction band. If light with an energy greater than the gap energy is incident on the crystal, a photon can be absorbed by an atom with the result that an electron can be excited into the conduction band, leaving behind the hole. In this way light creates electron-hole pairs in the crystal.

We are now in a position to understand how to construct a good solar cell. Light incident on the solar cell material will cause electron-hole pairs. We just need to collect the electrons (and holes) and direct them through an external circuit. To do this several important ingredients are necessary. The electrons and holes must move toward the external junctions of the solar cell without recombining in the process. This requires a cell region that is much like a perfect crystal, without

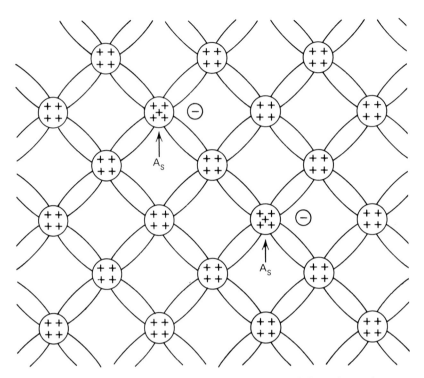

FIGURE 11-5 The addition of a small amount of arsenic with five valence electrons to the crystal structure of silicon forms what is called an *n*-type semiconductor. One electron is fairly free to move about while the stationary arsenic atom has a net positive charge.

impurities that would recombine with the electrons or holes. The material must be a good insulator, without a large number of free-charge carriers; otherwise the additional electrons and holes produced by the incident solar radiation will cause only an insignificant change in the total and thus would not be noticed. Also, a net electric field must be impressed on the solar cell in order that the electrons and holes be swept out of the generation region. The way to do this involves a scheme in which a pure silicon crystal is first made a semiconductor and then, later on, a region of very low conductivity with the required properties is obtained. This procedure is described below.

Pure silicon is not in a form that lends itself readily to absorb photons and give up electrons. To do this it is necessary to modify it by adding small amounts of impurities that greatly raise its electrical conductivity. Suppose an arsenic (As) atom is substituted for every millionth silicon atom. Arsenic has five valence electrons so that, since only four are needed for bonding to silicon, one electron is only weakly held by the positive charge of the nucleus. Such a "doped" crystal is called n-type silicon, and the bonding picture is shown in Figure 11-5.

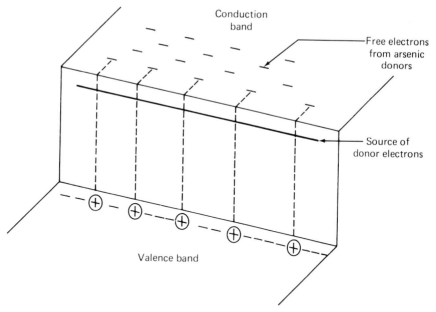

Conduction
band

Free electrons
from arsenic
donors

Source of
donor electrons

Valence band

FIGURE 11-6 Schematic representation of the band structure for *n*-type material. The extra free electrons in the conduction band come from the donor atoms with five valence electrons.

The excess positive charges in arsenic-doped silicon are bound, while the excess electrons move freely about. The net effect is to introduce an energy level near the top of the band gap that can provide electrons to the conduction band. The crystal as a whole is electrically neutral, but we now have a semiconductor with a few additional free electrons in the conduction band that can provide an electric current under the influence of an electrical field. The resulting band picture (in analogy with Figure 11-4) is shown in Figure 11-6.

It is also possible to form what is called *p*-type silicon by adding a small amount of atoms with only three valence electrons, such as boron (B). When a neutral boron atom occupies a lattice site in the silicon crystal structure, there will be one electron missing from the usual complete bonding. An entirely different electrical situation now prevails. A hole has been created in the lattice structure. Then, as described above, a neighboring electron can move over and fill the existing missing bond leaving behind a new missing bond and a net positive charge on a silicon lattice site. This process can occur repeatedly, with the result that electrons move in one direction (and the holes in the opposite direction). The motion of the holes can be crudely regarded as being equivalent to the motion of positive charges moving freely through the lattice. The movement of holes in the crystal lattice constitutes a flow of current. The addition of the boron has created an energy level near the valence band which can accept electrons and greatly increases the number of holes.

What happens now if light strikes a bit of semiconducting material? If the photon penetrates below the surface, and if its energy is equal to or greater than the amount of energy needed to move the electron into the conduction band, it will produce a conduction electron and its associated hole. Left by themselves the electron and the hole will wander randomly about the crystal lattice. If the electron encounters a hole, the two combine with the result that the energy that had been absorbed originally to create the electron and the hole is released as heat. To make a useful solar cell it is necessary to arrange things so that the electron passes through an external circuit, doing useful work, before it finally recombines with a hole. This can be accomplished by placing two slabs of semiconductor material together, with one being n-type and the other p-type.

The surface where the two types of silicon meet is called the p-n junction. Figure 11-7 shows a crystal in which the right-hand side has been doped with arsenic and the left-hand side with boron. Hence the right side contains a gas of free electrons and the left side a gas of free holes. Inside the junction region, diffusion causes a genuine, free-conducting electron to meet a true conducting positive hole; upon combining they neutralize each other. However, the bound positive

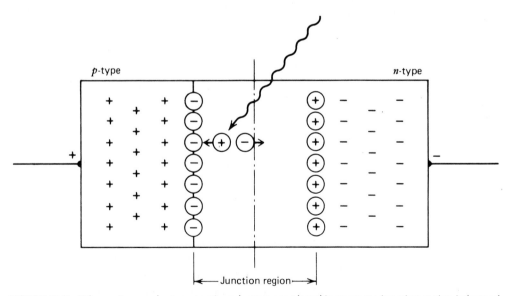

FIGURE 11-7 When p-type and n-type semiconductors are placed in contact a junction region is formed with practically no free charge carriers. An electric field is set up in the depletion region that sweeps out the electrons and holes (in opposite directions).

and negative charges are left, resulting in a potential difference between the two regions. The n-type side is positive with respect to the p-type side. This electrical field will prevent the electrons released by the incident solar radiation from dropping back into the holes from which they originated.

The solar cell works as follows. The incident solar radiation passes through the layer of p-type silicon and strikes the junction region. If absorption of light can be made to take place within a reasonable distance from the electric field of the p-n junction, the electrons will be drawn to the n-type material and the holes will be drawn to the p-type material. If the terminals to the cell are connected, current will flow in the external circuit as long as the junction is illuminated. The direction of the electron current in the external circuit will be from the contact on the n-type material to the p-type contact. For a practical cell it is important to have a large junction area so that the flat arrangement shown in the figure is used. Also, the p-type layer at the top should be as thin as possible in order to minimize nonuseful absorption of solar energy. A novel way of doing this is to diffuse boron at high temperatures into n-type silicon forming a thin-surface, p-type layer. The p-n junction is made over the entire sample so close to the surface that light can reach the barrier through the p-type material.

Is it possible to convert all of the energy of the incident solar energy into useful electric current? The answer is no, and the reasons for this can be demonstrated with the aid of Figure 11-8. The important loss mechanisms are shown on this figure along with their rough magnitude for a silicon solar cell.

In silicon the gap energy between the valence and conduction bands is close to 1.08 eV. Only photons with an energy greater than this can release electrons from the silicon crystal. This means that only light with a wavelength shorter than 1150 nm can release electrons. Roughly 23% of the incident solar radiation has a wavelength greater than this critical wavelength and hence cannot contribute to useful electric current.

Another loss mechanism is related to the fact that light with wavelengths shorter than 1150 nm will not only free an electron but also the excess energy over 1.08 eV goes into energy of motion of the electrons and is thereby eventually wasted as heat. Even photons of energy greater than 2 eV still release only one hole-electron pair each. It is now possible to make a crude estimate of the maximum energy available for conversion into electricity. The solar spectrum above 1150 nm contains 23% of the total solar energy incident and this is entirely lost. Below this region, a fraction equal to 1.08 eV divided by the energy is transformed into waste heat that only warms the solar cell. For example, the region of the solar spectrum between 500 to 700 nm contains 28% of the total energy, but only 15% can be converted into

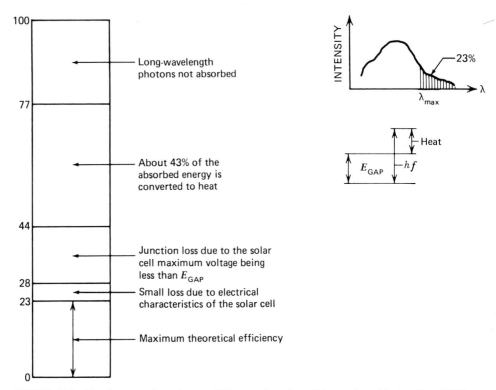

FIGURE 11-8 The theoretical maximum efficiency of a solar cell is considerably less than 100%.

useful electrical output. If this calculation is performed over the entire solar spectrum, it is found that the theoretical maximum efficiency for a perfect solar cell is now roughly 44%.

Any real solar cell has further losses. These have to do with the current-voltage operating characteristics of the solar cell along with some other electrical properties. The net result is that the net theoretical maximum efficiency is brought down to 28%. Finally, reflection of sunlight at the top surface must be taken into account. This effect can be up to 30%, but the loss can be substantially reduced by coating the upper surface with an antireflection coating. Thus, while the constraint of the second law of thermodynamics has been avoided, nature has replaced it with other restrictions making it possible to recover only a small fraction of incident solar energy.

It is evident that the temperature at which a photovoltaic device operates considerably affects the efficiency for energy conversion into electric current. The theoretical maximum efficiency for solar cells is shown in Figure 11-9 as a function of band gap, and for two different operating temperatures. Lines are drawn for three common types of cells—silicon, gallium-arsenide, and cadmium-sulfide. Notice that the maximum efficiency of a silicon solar cell drops from 25% at 0°C to

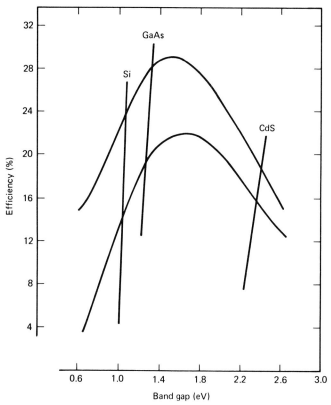

FIGURE 11-9 The maximum possible efficiency of a solar cell depends in a complicated way on the band gap of the semiconductor material and on the temperature.

14% at 100°C. It is reasonably easy to qualitatively understand the variation of efficiency with band gap. The curve goes up at first as the band gap increases since the thermal contributions to the number of electrons in the conduction band decrease. Later the decrease in the number of photons absorbed dominates and the efficiency decreases.

Now that we know how a solar cell responds to sunlight it will be useful to study how the voltage and current produced by the solar cell vary under different load conditions. Also, we need to look at the output current (or voltage) as a function of the intensity of the incident radiation.

The above factors can be illustrated using the curves of Figure 11-10. Look first at the topmost curve which corresponds to 1000 w/m² (often called one sun). This curve is obtained by putting the solar cell into a circuit with a variable load resistance and measuring the voltage across the cell and the current through the load resistance. When the load resistance is very high (say an open circuit) the current will be zero and the voltage will be at its maximum. This latter corresponds to about

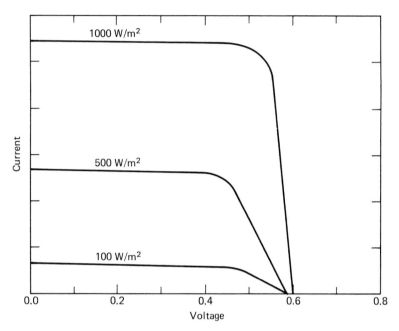

FIGURE 11-10 Current-voltage characteristics of a solar cell. The current units are left arbitrary. The solar cell output is shown for three different amounts of sunlight.

0.50 V for a silicon solar cell. As the load resistance is decreased the current rapidly increases and approaches a maximum when the voltage is about 0.45 V. As the load resistance is further decreased, the current remains constant while the voltage decreases.

In practical applications one usually operates the solar cell near the knee of the curve. This is done by adjusting the load resistance to the proper value. Operating the solar cell at this point provides the maximum transfer of power to the load.

The remaining curves correspond to the solar cell characteristics as the light intensity is varied. The shape of the current-voltage curves remains almost the same as the light decreases. The current, however, decreases linearly with light intensity. This property is useful in many applications.

B.   CRYSTALLINE SILICON

The majority of solar cells are produced as circular wafers from single-crystal silicon. Every effort is made to fabricate useful p-n junctions in as simple and low-cost a method as possible. Nevertheless, the present cost of silicon solar cells is high. In this section we discuss how single-crystal solar cells are made and also study a few of the attractive possibilities for lowering their cost. Amorphous-silicon solar cells, which avoid the need for crystalline silicon, are discussed in the next section.

It is worthwhile to note a few of the important constraints that must be borne in mind in fabricating solar cells. Solar cells need to be fairly large in area (commonly 10-cm-diameter wafers) with uniform electrical characteristics. The amount of this surface area exposed to the sunlight must be maximized. A typical solar cell will have about 90% active cell surface. In addition, good electrical contact to the top and bottom surfaces must be made, so that resistive losses in the contacts are minimal. Finally, the incident energy deposited as heat must be carried away. Otherwise, the temperature of the solar cell would rise with a resultant decrease in conversion efficiency.

The manufacturing process used today is primarily responsible for the high price of solar cells. Briefly, the process presently used to manufacture silicon solar cells is as follows. The jumbled, polycrystalline structure of single-crystal silicon is first converted into an orderly structure of single-crystal silicon by what is known as the Czochralski growing process. This provides single-crystal silicon in the form of cylindrical ingots two or more inches in diameter. To prepare an ingot, a batch of silicon is melted and minute amounts of boron or phosphorus added. Then, a silicon seed crystal is lowered into the 1420°C liquid, and an ingot is "grown" by twirling and pulling the seed upward at a few inches per hour. If the temperature and the rates of the rotation and pulling are controlled carefully, a perfect crystal will be formed. Once an ingot is grown, up to three-quarters of it is destroyed in the remaining production process. A circular diamond saw cuts thin slices about 25 to 30 thousandths of a centimeter thick. Since the saw blade is also this thick, much of the ingot becomes sawdust, thereby raising the cost. Grinding and polishing these silicon wafers pushes labor costs higher. The wafers are then baked in a chemical atmosphere that diffuses another semiconducting layer into one surface. Finally, electrical terminals are attached, and the wafers are wired into solar panels. The reason for the high cost of solar cells is that the growth, slicing, and fabrication processes are all highly labor intensive, requiring skilled workers. Automating some steps should bring costs down to less than $5 per watt in a few years, which is still too expensive for large arrays.

From the above description of the fabrication process for single-crystal silicon solar cells, it is apparent why their present cost is so high. We can briefly summarize the key factors contributing to this cost as follows:

1. Need for high-purity starting material.

2. Large loss of material during fabrication.

3. Slowness of the crystal growth and diffusion processes.

4. Energy and labor intensive methods needed for production and fabrication.

Many different schemes have been suggested to lower the cost of silicon solar cells. These include lower-cost starting material and doping by ion inplantation.

One particularly promising way to lower the crystal-growing cost involves the growth of continuous ribbons of silicon by a process known as edge-defined, film-fed growth (EFG). Originally, EFG was developed to speed the growth of single-crystal sapphire used for electrical microcircuits. When a die is lowered into molten silicon or sapphire material, the liquid rises through its center by capillary action. The material flows only to the top edge of the die, whose shape determines whether a tube, square, ribbons, etc. is formed. A seed crystal is placed in this liquid film. As the seed is carefully drawn upward, it grows into a thin ribbon of silicon crystal whose width and thickness are determined by the dimensions of the die. The ribbon can be cut into wafers with a minimum of waste since there is no need for sawing or polishing. Ribbons several feet in length have been pulled from a crucible of molten silicon in one hour. A schematic illustration of the EFG process is given in Figure 11-11.

Fabrication of the n-type region of each wafer is done as usual by diffusing phosphorous into the front surface. The remaining steps in the solar-cell fabrication process involve contact metallization and application of an antireflection coating. The best antireflection coatings are vacuum-deposited layers of titanium oxide or tantalum oxide. While material costs are low, production costs are very high and better methods are under study.

The silicon crystals grown by this EFG process are structurally far less perfect than those grown by the Czochralski method. Solar cells with efficiencies between 10 to 12% have been constructed from this EFG material, but the process introduces unwanted impurities into the silicon and is not yet as rapid as the traditional method.

The best way to reduce the costs of silicon solar cells is to relax the single-crystal requirement. An attractive way to do this is to make solar cells from thin layers of polycrystalline silicon rather than from single-crystal wafers. This would avoid the expense of growing and preparing very pure wafers of single-crystal silicon and at the same time would greatly reduce the amount of silicon material needed.

Polycrystalline silicon is comprised of many small silicon crystals called grains, which are oriented essentially randomly. The boundary between adjacent grains is a region of disorder. Making solar cells from this type of material has proven difficult because the presence of grain

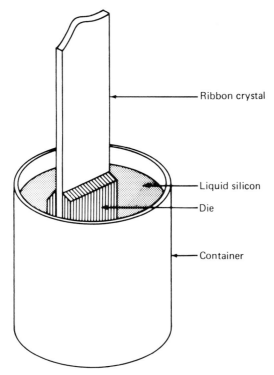

Ribbon crystal

Liquid silicon

Die

Container

FIGURE 11-11   Schematic drawing of the EFG crystal-growing process. As described in the text, a ribbon of crystal silicon is obtained continuously from the liquid silicon.

boundaries near and parallel to the p-n junction results in a reduction in the effective thickness of the solar cell. One way to overcome this problem is to make the grains fibrous in shape. Efficiencies of 5 to 7% have been reported for some types of polycrystalline solar cells.

C.   AMORPHOUS SILICON

The high cost of most solar cells derives from the need to use highly refined single-crystal silicon as the primary constituent. Perhaps the most promising way around this is through the use of amorphous silicon, where no regular crystal structure exists. In this approach solar cells are made by depositing a thin film of amorphous silicon onto a substrate (backing material). This approach has three useful advantages:

1.   The production and fabrication processes would be much simpler and cheaper.

2. The process, involving deposition of silicon, lends itself readily to mass production.

3. The demand for silicon starting material is greatly reduced. Thin-film amorphous silicon cells are approximately 10^{-6} m thick, whereas single-crystal silicon cells are orders of magnitude thicker.

The explanation of why amorphous silicon can be so thin is somewhat complex, but is very worthwhile since it involves many of the important concepts about photovoltaic devices.

We saw earlier that a perfect crystal has a forbidden region in which no electrons are present. This is illustrated again in Figure 11-12, in which the language has been slightly changed. Instead of talking about electrons we speak of the density of states as a function of electron energy. When the density of states is low, very few electrons are present. Pure amorphous silicon is formed by the condensation of silicon vapor onto a cold substrate. The resulting silicon material does not have the ordered (crystalline) structure found in single-crystal material. Another problem with amorphous silicon is that the nice sharing of electrons that occurs in a regular crystal is not present. This results in what is termed *dangling bonds*. Because of these problems,

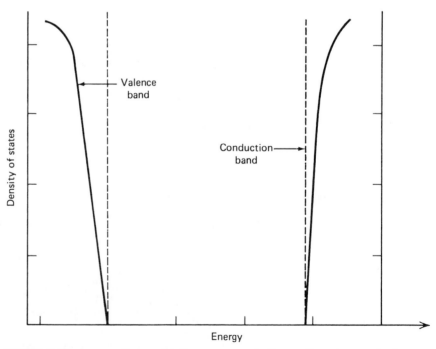

FIGURE 11-12 In a perfect crystal there are practically no allowed states in the gap region.

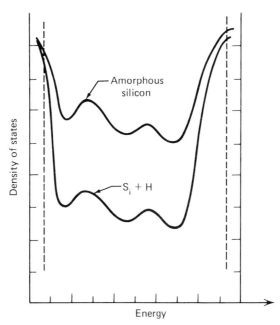

FIGURE 11-13 Density of states in the forbidden gap of amorphous silicon. Note the striking improvement due to doping with hydrogen.

many allowed energy states become possible within the normally forbidden band gap between conduction and valence electrons. This situation is pictorially illustrated in the top portion of Figure 11-13.

In 1976 scientists discovered that doping amorphous silicon with hydrogen greatly reduced the number of states in the forbidden region. One way to do this is to pass silane gas (SiH_4) through a radio frequency coil and let the gas impinge on a cold substrate. This process puts hydrogen into the material in such a way that some of the dangling bonds are satisfied. The dramatic decrease in the density of states in the gap region is shown in the lower curve of Figure 11-13.

Two other important results have come from the research and development work on amorphous silicon solar cells. First, it was found that amorphous silicon could be doped, that is, p- and n-type material could be formed to produce a semiconductor junction. Also, it was found that the absorption coefficient of amorphous silicon for visible light is over 10 times that holding for single-crystal silicon. In other words, a significant portion of the incident solar intensity could be absorbed in a thin film only 10^{-6} m thick.

Early amorphous silicon solar cells had an efficiency of only 1 or 2%. But, considerable improvement has occurred over the past few years.

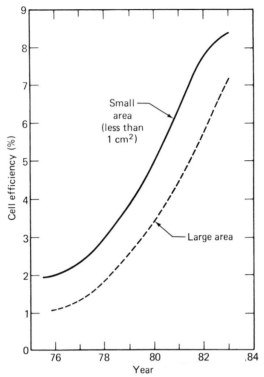

FIGURE 11-14 The increase in efficiency of amorphous silicon solar cells with time is depicted. Large-area cells with an area of 100 cm² and an efficiency of over 6% have been constructed. (Source of data is the Solar Energy Research Institute.)

The historical trend of cell efficiency with time is shown in Figure 11-14. The lower curve corresponds to solar cells with an area greater than 1 cm². The theoretical upper limit on the efficiency of amorphous silicon solar cells is estimated to be between 15 to 20%. With so much at stake it appears certain that this promising direction will continue to be actively pursued.

D. OTHER APPROACHES

Solar cells made from silicon are not the only possibility. Cells made from thin films of cadmium sulfide (combined with copper sulfide and other compounds) have been intensively studied, particularly at the Institute of Energy Conversion at the University of Delaware. The thickness of these cells is usually in the range of 5 to 50 millionths of a meter.

A common method of fabrication is to plate a metal or plastic substrate with zinc and then evaporate a layer of cadmium sulfide onto the substrate. The next steps involve etching, dipping into an aqueous solution of cuprous chloride, adding a contact grid to the top, and finally encapsulating the entire unit in a protective transparent covering. Efficiencies in the range of 5 to 9% have been readily achieved.

The way to achieve rapid commercialization of photovoltaic systems is obviously to decrease costs. Another excellent possibility to reduce costs is to reduce the amount of expensive solar cell material needed. This can be achieved by using some type of concentration system to focus the sunlight collected over a large expanse of land onto a much smaller area of solar cells.

Systems presently under consideration cover a range of concentrations from 10 to over 1000. A variety of innovative concentrating collectors have been designed. With one exception, the concentrating systems will be sun tracking. The exception is the trough-shaped Winston collector. This device has the nice property of focusing all incident radiation within a certain range of angles on a small amount of surface area. A concentration factor of about 10 can be utilized without the need to track the sun across the sky. At high concentrations most collectors will require cooling, because the performance of solar cells goes down as the temperature of the device increases. A tremendous advantage of concentrating systems is that economic studies show that the cost of generating electric power is almost independent of the cost of solar cells, provided that the latter is below a critical value of about $1000 per square meter.

Reflectors used as concentrators are fabricated as troughs or dishes. The trough geometry has been the most popular choice for photovoltaic concentrators. This type of concentrator is particularly suited to the circulation of a cooling fluid. Also, trough units can be tracked in many different ways. One common trough approach is illustrated in Figure 11-15.

Another concentrating method involves the use of a refracting lens. Ordinarily this approach is not used because of the loss of solar energy in passing through the lens material. For solar cells that are situated directly behind a plastic lens, this problem is not so severe. An example of a prototype fresnel-lens concentrator constructed by Sandia Laboratories is shown in Figure 11-16. Each fresnel lens, which concentrates the sun's rays like a magnifying glass, focuses the equivalent of 50 suns on a corresponding solar cell to convert sunlight directly to electricity. The high-intensity illumination increases each solar cell's electrical output, thereby reducing the resultant cost of the solar electricity.

It is generally felt that concentrating photovoltaic systems will need to utilize very high-efficiency solar cells. Because of this, interest is high in gallium-arsenide photovoltaic cells, despite the fact that they

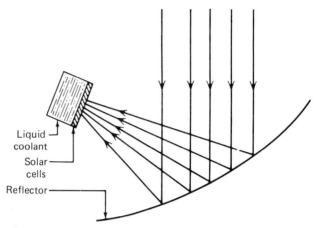

FIGURE 11-15   Parabolic trough-type concentrating device.

are presently 10 times as expensive as their silicon counterparts. Varian Corporation has demonstrated an experimental system with an efficiency of 19% when operated with a concentration factor of over a thousand, while IBM has tested a gallium-arsenide cell having an efficiency of 22%. A key advantage of gallium-arsenide cells for this

FIGURE 11-16   Using 135 plastic fresnel lenses mounted directly in front of the corresponding number of silicon solar cells, the photovoltaic concentrator array pictured above produces 1 kW of electricity. (Courtesy of Sandia Laboratories, Albuquerque, New Mexico.)

application is that they can tolerate higher temperatures than silicon, up to 200°C with only a moderate loss in efficiency. Still to be answered is the question of whether or not a sufficient resource of gallium is available to permit the large-scale use of gallium-arsenide solar cells.

If the cost of solar cells continues to decrease as predicted, both small and large arrays of photovoltaic devices will soon become a practical eventuality. One very intriguing usage of large solar-cell arrays is to place them in orbit above the earth. In this case the problems of irregular collection due to clouds would be eliminated. Storage of energy would no longer be a problem, since a single satellite would view the sun for 23 out of every 24 hours. The intense, almost constant solar flux reduces the collector area requirement by almost a factor of 8. It is easy to calculate the area required in space needed to provide the U.S. energy consumption of 1970 (6×10^{19} J) assuming an overall conversion efficiency to electrical power of 15%. The answer is that an area of about 9500 km^2 would be needed, which is one-eighth as much as would be needed for a surface location. On a smaller scale, the area needed for a 1000-MW plant would be under 5 km^2.

Spurred on perhaps by the success of the U.S. satellite program, Leon Gauderer of Texaco made the first serious suggestion for the use of an orbiting power satellite in 1965. In 1968, Peter Glaser of the Arthur D. Little Company suggested a detailed proposal for the large-scale production of power in space. A large satellite, such as that shown in Figure 11-17 would be placed into a synchronous orbit of the earth, remaining fixed above a given ground location. The satellite would consist of a solar collector, an energy conversion device to provide electric power, a transmission cable to supply the electricity to micro-wave generators, and an antenna arrangement to beam the microwave radiation to a receiving station on earth.

The geometrical arrangement for the satellite system is shown in Figure 11-18. At an altitude of 35,500 km in an orbit parallel to the earth's equatorial plane, a satellite moving from east to west would be stationary with respect to any given point on earth. Since the satellite will lie in the earth's shadow once a day, two satellites in the same orbit, but spaced 21° apart, would be required in order that at least one would be illuminated at all times. A network of satellites that could continuously service a large number of stations on the ground has been suggested.

While the develoment of economic photovoltaic devices is the single-most important technical development needed for this satellite approach to succeed, there are many other highly interesting techno-logical problems that would have to be overcome. For example, the solar collectors should be oriented toward the sun to within about 1°. Glaser maintains that, except for system size, this is within the realm of present tracking technology. A more difficult requirement is that of

FIGURE 11-17   Solar electricity could be generated in space by mounting huge arrays of solar cells on an orbiting synchronous satellite. The collected energy would then be beamed back to earth in the form of microwave radiation. (Courtesy of Arthur D. Little, Inc., Cambridge, Mass.)

locking onto an earth-based antenna. In order to keep the microwave beam from straying more than about 150 m, a pointing accuracy of 0.5 sec of arc is required. A guidance and control system will have to neutralize the effects of various forces acting on the satellite in order to maintain a stable orbit. The largest of these forces is due to radiation pressure from sunlight; small control rockets may be used to help

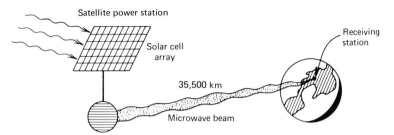

FIGURE 11-18 Illustration of the solar satellite concept.

overcome this. Cooling of the microwave generators will have to be handled by some type of closed circulating system, with the waste heat being radiated into space by means of heat pipes on space radiators that are distributed over the structure of the microwave antenna.

Environmental problems are a serious consideration for this approach. A power density of about 1 W/cm^2 for the transmitted radiation has been suggested. This is roughly an order of magnitude higher than the normal solar radiation intensity. It is sufficiently high to damage objects or living tissues that enter the beam. Regulations and controls to prevent the entry of objects or living beings into the beam would have to be imposed. Another serious problem involves the use of a space shuttle to set up the solar satellite system. Would continued passage of space shuttles through the ozone layer break down this vital safeguard against ultraviolet radiation? And, of course, the cost of the many space shuttle flights required to transport the needed materials into space may turn out to be prohibitive.

REFERENCES

1. D. M. Chapin, "The Direct Conversion of Solar Energy to Electrical Energy," Chapter 8, of *Introduction to the Utilization of Solar Energy*, A. M. Zarem and D. D. Erway, eds. (New York: McGraw-Hill, 1963).

    A review of semiconductor principles—design of solar cells, their fabrication, theoretical efficiency, and economics.

2. B. Chalmers, "The Photovoltaic Generation of Electricity," *Scientific American* (October 1976).

    An elementary and very readable discussion of the principles of solar cells. Contains a useful summary of present manufacturing methods.

3. *Sunworld*, Volume 4, Number 1, 1980 (New York: Pergamon Press).

    This entire issue is devoted to the subject of photovoltaic devices. There are two excellent articles on amorphous silicon solar cells.

4. *Solarex Guide to Solar Electricity* (Solarex Corporation, 1979).
    A simple, concise treatise about solar cells.

5. Martin A. Green, *Solar Cells* (Englewood Cliffs, N. J.: Prentice-Hall, 1982).
    An excellent technical treatment of the subject for the advanced reader.

QUESTIONS

1. n-type silicon is formed by adding minute amounts of elements with the following number of valence electrons:

    (a) 3   (c) 5
    (b) 4   (d) 6

2. At room temperatures the theoretical maximum efficiency of a solar cell cannot exceed about:

    (a) 5%   (c) 45%
    (b) 25%   (d) 75%

The next 9 questions refer to the figure below.

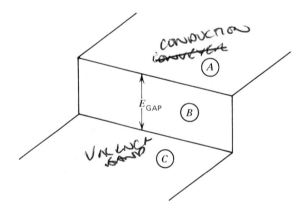

3. What is region *A* called?

4. What is region *C* called?

5. What is the difference between a conductor and a silicon crystal as far as region *A* is concerned? (*Hint.* Discuss the number of electrons present.)

6. Explain why light of very long wavelengths cannot cause any current to flow in a silicon solar cell.

7. If an electron is elevated to region *A* from region *C*, what do we call the absence of an electron in region *C*?

8. Solar cells are an extremely attractive way to generate electricity. What is the most important reason why this energy source is not yet being utilized on a large scale?

9. What is meant by "n-type" silicon?

10. p-type silicon has an excess of what type of charge carrier?

11. The incident light waves can also be thought of as "photons" of energy hf. What can you say about the energy hf as compared to E_{GAP}. $hf > GAP$

12. Name at least two major difficulties with the production of single-crystal solar cells.

13. In n-type silicon are there any charge carriers in the conduction band in the absence of sunlight?

14. Can you suggest something besides arsenic as the impurity for making n-type-silicon, semiconductor material.

15. Do the same thing as in Problem 14 for p-type-silicon, semiconductor material. This time you are looking for a substitute for boron.

16. Why does one often construct solar cells in the form of a flat wafer?

17. Explain why the efficiency curves of Figure 11-5 fall off at high-band-gap energies.

18. Do practical solar cells need to be cooled? Explain.

19. In the discussion of the plan to put a large array of solar cells out in space, it was noted that the required surface area in space for the solar cells would be almost a factor of 8 less than for a similar power plant on earth. Explain how this factor of 8 is obtained.

20. Solar cells are one of the many applications of solid-state devices. In this connection it is worth noting that transistorized electronics are much more reliable than their vacuum-tube counterparts. List the reasons why this statement is true.

21. What is the key advantage of polycrystalline silicon solar cells over those grown from single crystals?

22. What is the advantage of using concentrators with solar cells?

23. Name at least three factors responsible for the low efficiency of solar cells.

24. What is meant by the "forbidden" band gap?

25. Give a reason why the addition of hydrogen to amorphous silicon improves the cell efficiency.

The following questions refer to the figure below.

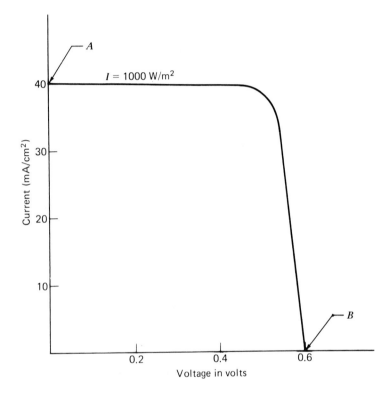

26. What is the open-circuit voltage?

27. What is the short-circuit current?

28. What load resistance is needed to operate at the "knee" of the curve?

29. What will be the cell current when the incident solar flux is 400 W/m^2 and the output voltage is 0.40 V?

30. What is the power delivered at point A?

31. What is the power delivered at point B?

appendix I

---

flat-plate collector heat loss analysis

In Chapter 7 it was stated that the heat loss through the top of a flat-plate collector is

$$Q = U_{top}\Delta T \tag{I-1}$$

where U_{top} is an overall upward heat transfer coefficient and ΔT is the temperature difference between the absorber plate and the outside. The absorber plate temperature can be approximated by the average of the inlet and outlet temperatures of the heat transfer medium. (Correction must also be made for the nonuniform nature of the absorber plate temperature distribution.) The purpose of this presentation is to illustrate how U_{top} can be obtained from the individual heat transfer coefficients corresponding to the upward heat losses between adjacent surfaces, and to show the dramatic decrease in heat loss as the number of cover plates is increased.

The procedure will be outlined for a liquid-type collector as shown in Figure I-1. The total temperature difference between the absorber plate and the outside world can be written as

$$\Delta T = \Delta T_1 + \Delta T_2 + \Delta T_3 \tag{I-2}$$

At equilibrium, the heat loss Q between adjoining surfaces is given as

$$Q = U_1\Delta T_1 = U_2\Delta T_2 = U_3\Delta T_3 \tag{I-3}$$

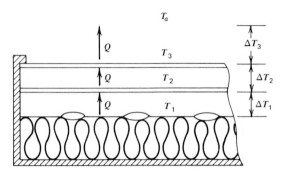

FIGURE I-1 The temperature distribution in a flat-plate collector with two cover plates.

In this relation, each of the U's is equal to the sum of its radiative and convective components. If we substitute for the ΔT's, we have

$$\Delta T = \frac{Q}{U_1} + \frac{Q}{U_2} + \frac{Q}{U_3} = Q\left(\frac{1}{U_1} + \frac{1}{U_2} + \frac{1}{U_3}\right) = \frac{Q}{U_{\text{top}}} \qquad \text{(I-4)}$$

where

$$\frac{1}{U_{\text{top}}} = \frac{1}{U_1} + \frac{1}{U_2} + \frac{1}{U_3} \qquad \text{(I-5)}$$

The calculation of U_{top} is thus easily done once the various individual terms are known.

To conclude this discussion it is instructive to illustrate how additional cover plates reduce the overall heat loss. Suppose that without any cover plates, the heat transfer coefficient for the top is 30 W/m²-°C, a typical value for an unprotected surface in a mild wind (which causes severe convective losses). The heat loss Q would then be 30 ΔT. If a cover plate is added, the heat transfer coefficient between the absorber plate and the glazing is now about 9 W/m²-°C, while the heat transfer coefficient between the glass cover plate and the outside world is still 30 W/m²-°C. The overall heat transfer coefficient is

$$\frac{1}{U_{\text{top}}^1} = \frac{1}{30} + \frac{1}{9} = \frac{1}{6.92}$$

The heat loss from the collector is now 6.92ΔT, a reduction of over a factor of 4.

The addition of a third cover plate gives

$$\frac{1}{U_{\text{top}}^2} = \frac{1}{9} + \frac{1}{9} + \frac{1}{30} = \frac{1}{3.91}$$

The upward heat loss is now reduced to $3.91\Delta T$, which is a further substantial reduction over the heat loss obtained with only one cover plate. Remember, of course, that this reduction in heat loss is accompanied by a loss in the overall transmission through the additional cover plates. At higher temperatures the reduction in heat loss normally justifies the use of two cover plates.

answers to questions

Chapter 1

1. b
2. d
3. b
4. b
5. b
6. c
7. b

8. d
9. a
10. c
11. a
12. c
13. b

14. (a) Yes
    (b) No
    (c) No
    (d) Yes
    (e) No
    (f) Yes
    (g) No
    (h) Yes
    (i) No
    (j) Yes

15. It appears that the amount of useful energy obtained from a barrel of shale-rock oil is slightly less than the amount of energy needed to manufacture that barrel of oil.

17. It will, in principle, take an infinite amount of time.

18. a, c, e are all possible causes of the energy crisis.

19. 1. g
    2. i
    3. c
    4. b
    5. e
    6. a

21. Obtain the monthly average for the summer electrical consumption. Subtract this from the monthly winter electric consumption to obtain the space heating portion.

22. (a) Increased humidity in the building.
    (b) Buildup of pollutants in the house.

23. The heat from the lights goes into useful space heating during the winter.

24. For example, this reduces the amount of litter on the roads.

25. Fiberglass cuts down on convection through the material. Putting insulation on the outside of a concrete wall allows the thermal mass of the concrete to be used to store energy and smooth out temperature variations in the building.

27. Cooler in the summer, by reducing the heat gain through the top of the house.

30. d

Chapter 2

1.	a	**8.**	a
2.	b	**9.**	b
3.	c	**10.**	d
4.	c	**11.**	c
5.	c	**12.**	b
6.	b,c,f	**13.**	c
7.	b		

14. Watt, J/sec. Btu/hr, horsepower

15. It is released into the outside environment. No.

16. 240 kJ

17. No. The efficiency of the electric generating plant is only about 30%, while that of the furnace is perhaps greater than 70% (for space heating).

18. The greater mass of the heavy air implies that a larger amount of force will be required to move it (in most cases).

20.	1000 MW		
21.	d	**24.**	a
22.	c	**25.**	a
23.	a	**26.**	d

27. It increases, as work is done on the gas.

28. It decreases.

29. Before.

30. 90% or more depending on the efficiency of the wind turbine. Ordered.

31. The working fluid is brought back to its original state.

Chapter 3

1.	a	**8.**	a
2.	b	**9.**	d
3.	c	**10.**	c
4.	d	**11.**	c
5.	c	**12.**	b
6.	c	**13.**	a
7.	b	**14.**	b

15. ^{235}U

16. Colorado, Utah, Wyoming.

17. i, o, b, l, g, f, m, r.

18. Because the ordinary water absorbs neutrons. This means that more fissionable ^{235}U atoms are needed to make the reactor go.

19. Water.

20. The North Sea.

21. Shale rock, tar sands, coal.

22. The production of synthetic (power) gas.

23. The need for reclamation of the land that has been strip-mined.

24. The addition of a slow neutron to ^{235}U causes the resulting nucleus to become unstable and to split into two parts with the subsequent release of a large amount of energy.

25. Hot, dry steam.

26. More fissionable material is produced in the reactor than is consumed.

27. Heavy water (D_2O).

28. (1) Low water temperature means low efficiency of conversion to electric power. (2) There is a large quantity of pollutants in the hot water.

29. By converting the coal to either gas or some form of petroleum.

30. Too difficult and expensive to extract from the shale rock. (See also the answer to Question 28.)

31. (1) Solar energy, (2) nuclear fusion, (3) nuclear fission.

32. 6%.

33. About four years.

34. No, it would just transfer the pollution problem to the central power plant.

35. About 25%.

36. Because of the relatively low temperature of the hot wet steam or hot water.

37. 16%.

38. Because they are geologically stable.

39. d

40. a

41. c

42. No. One neutron (at least) is required to make another fissionable ^{239}Pu atom. The 1.0 neutrons left over are not sufficient to run a fission reactor, as nothing is left to allow for neutrons lost due to leakage, absorption, and so on.

Chapter 4

1. c

2. a

3. c

4. d

5. b

6. Tax credits.

7. F. Shuman.

8. H. Thomason.

9. Metal shields.

10. 20 years.

11. New energy is saved. The private and public power companies are then able to use their old energy sources for new purposes. New energy is much more expensive than using older sources such as dams and power plants constructed many years ago. Provision should be made to compensate the solar installer for this difference.

Chapter 5

1. a	**4.** c	**7.** b	**10.** a	**13.** a
2. c	**5.** c	**8.** a	**11.** b	**14.** b
3. b	**6.** c	**9.** b	**12.** a	

15. 1-d, 2-b, 3-c, 4-a.
16. (a) Roughly 5 billion years.   (g) 5800 K.
(b) Helium.   (h) 11 years.
(c) Greater than.   (i) General relativity.
(d) Different.
(e) Sunspots.
(f) One million years.
17. Because the first step in the basic process

$$p + p \rightarrow D + \beta^+ + v$$

goes because of the weak interaction. This reaction proceeds too slowly to be seen on earth.
18. The orbital motion around the center of the galaxy requires the gravitational force toward the center. Hence, the sun cannot collapse toward the galactic center.
19. Zero.
20. a, d, f.
21. (1) Sun's power level (in the core) is way down. (2) Neutrinos decay on the way from the sun. (3) Something is wrong with our understanding of the basic reaction processes.
22. They are located in our own galaxy.
24. Three times as fast.
25. The lighter elements boiled off the earth, which is not massive enough to hold on to hydrogen, etc.

Chapter 6

1. b	**5.** d	**9.** d	**13.** a	**17.** a	**21.** c
2. b	**6.** a	**10.** c	**14.** a	**18.** c	**22.** c
3. a	**7.** c	**11.** a	**15.** c	**19.** a	
4. a	**8.** a	**12.** a	**16.** b	**20.** d	

23. a) Tranverse, longitudinal.
b) Transverse.
c) 20 cm.

d) $V = f\lambda$
e) $f = 5/\text{sec}$
f) 10 cm.
g) The electric and magnetic fields.
h) The water molecules.

24.   1 j   5 l
      2 e   6 c
      3 d or j   7 k
      4 i   8 a

25. Water vapor.
26. On cloudy days the solar radiation is mostly diffuse.
27. It is converted to thermal energy of the molecules in the metal bar.
28. (a)   8%
     (b)   84%
     (c)   90%
     (d)   94%
29.   1. Empire State Building.
      2. Wavelength of radio waves.
      3. Velocity of light in air.
      4. Amount of light reflected by a mirror.
      5. Amount of light reflected by a pane of glass with an angle of
          incidence equal to 60°.
      6. Frequency of a light wave.
30. The image is perverted, so that the right hand appears to be the
     left, etc.
31. The sun is behind you, and the rain drops are in front.
32. Critical angle occurs when light is totally reflected; will occur for
     light coming out of water for θ_i less than 90°.

Chapter 7   1. To store excess heat available during periods of high insolation.
                           This energy is given back to the living space during periods of low
                           or no insolation. This process helps avoid large-temperature
                           savings in the living area.
                       2. (a)   North-south orientation.
                           (b)   Large thermal mass.
                           (c)   Overhangs for summer protection.
                           (d)   Movable insulation over glazings.
                           (e)   Large expanse of south facing glass, etc.
                       3. To provide extra insulation and cut down on building heat losses.
                       4. Because of convection, heat is quickly transported to the back face
                           of the storage wall. This process is independent of the wall
                           thickness.
                       5. b.
                       6. Roughly 8 to 16 in.
                       7. By going to a thicker wall, temperature fluctuations in the living

area are smoothed out. This increase in comfort more than compensates for the small loss in annual solar energy collected by the system.

8. The heat loss, principally at night, is quite large with single glazing.

9. Yes. This prevents warm air from the interior of the house from coming in contact with the cooler front glazing—with a resultant heat loss.

10. It is painted black in order to absorb as much of the incident solar radiation as is possible.

11. (a) By convection from the front wall to the interior of the house.
    (b) Heat is also conducted from the front face to the back face, where it is then transported to the living space by radiation and convection.

12. 120°F.

13. Yes; see answer to question 11.

14. Put in a vent at the top of the glazing.

15. One would choose the 365-ft^2 wall. This would give 60% of the required space heating. To obtain 80% of the building's space heating needs would mean an increase in wall size (and proportionate cost) of about 2.3.

16. Single-glazed system.

17. 95°F.

18. (a) Glass.
    (b) Glass.
    (c) Glass.

19. The roof-pond insulation is kept down during the day and raised at night.

20. (a) Greenhouse.
    (b) Trombe wall.
    (c) Double-envelope building.
    (d) Superinsulated house.
    (e) Direct-gain structures.
    (f) Roof-ponds.

21. B.

22. B.

23. The fraction of the total winter heating load that is obtained from the solar system.

24. 1.00.

25. Because this would result in the solar system providing too much heat during portions of the fall and spring.

26. Superinsulation.

27. Concrete, masonry, and water.

28. There must be too much direct gain in this particular room.

Chapter 8

1. a	**6.** c	**11.** c	**16.** c	**21.** a
2. c	**7.** a	**12.** a	**17.** b	**22.** c
3. d	**8.** d	**13.** c	**18.** a	
4. c	**9.** a	**14.** c	**19.** d	
5. c	**10.** b	**15.** c	**20.** d	

23. $Q_{rad} = \varepsilon\sigma(T^4 - T_1^4)$

24. It will increase.

25. Because there must be a net heat flow from the absorber plate to the cover plate, T_2 is less than T. Also, since there must be a net heat flow from the cover plate to the outside world, T_2 must be greater than T_a.

26. c

27. B

28. A

29. The temperature difference between the absorber plate and the outside world.

30. About 0.82.

31. The trickle-type collector.

32. An evacuated tube collector.

33. An evacuated tube collector.

34. Because the reduction in heat losses in going from two to three cover plates is usually less than the increased reflection and absorption loss in the third cover plate.

35. Convective heat loss.

36. a	**39.** c	**42.** c
37. b	**40.** a	**43.** b
38. c	**41.** a	**44.** c

45.

1 j	4 b	7 m
2 e	5 c	8 a
3 p	6 o	9 k

46. It is less than the ideal collector's efficiency.

47. About 90° from the horizontal (almost vertical).

48. It decreases.

49. Crab (or Winston) collector.

50. A parabolic surface.

51. 10.

52. No, it is nonimaging.

53. It will be captured by the system.

Chapter 9

1. c	**5.** c	**9.** a
2. b	**6.** d	**10.** b
3. c	**7.** d	
4. b	**8.** b	

11. 38°C.

12. It is the amount of heat required to raise unit mass of material one degree of temperature.
13. Curve 1, the heat-of-fusion salt.
14. The storage of heat by a material by raising its temperature.
15. c   18. d
16. a   19. b
17. c
20. Rocks, or gravel.
21. This ensures that the water entering the collector will be heated to a temperature above that of the tank. At the same time, the 10 to 15°F differential helps avoid unnecessary cycling of the system.
22. To maximize the surface to volume ratio.
23. Drainback, draindown, recirculation, antifreeze, and electrical resistance heating are examples of freeze-protection approaches.
24. (a) Air leaks are hard to detect.
    (b) Lower collector efficiency (under same operating conditions as a liquid system).
    (c) Rock storage has lower heat capacity per unit weight.
    (d) Need special size rocks (not too large or too small).
25. (a) No freeze protection worries.
    (b) No bothersome water leaks.
    (c) Easy to handle.
    (d) Good heat transfer from the storage system to the house (better than is achieved with liquid systems).
26. Mass and difference of temperature.
27. Water, rock, and heat-of-fusion salts.
28. Because it also collects the diffuse radiation.
29. Five pounds of water.
30. (a) They suffer from "supercooling."
    (b) They can be corrosive.
    (c) They form clumps of material that will no longer go through the phase-change process (after many cycles of operation).
31. The water drains out of the system into the small reservoir. (This is the drainback system).
32. It contains the water in the closed loop when the solar system is not in operation.
33. About 80 gal.
34. The income tax credit of 40%.
35. Whenever the collector pump is turned off, the water in the solar collector is drained out through the drain valve (to the ground, sewer, etc.).
36. Draindown.
37. The fluid in the collector is drained away each time the solar collector is turned off. This technique would quickly deplete the circulating fluid in a closed-loop system.

38. The water at the bottom of the storage tank is cooler. This means that the average temperature of the solar collector will be lower than if water from the top of the tank were used. Operating this way, heat losses are reduced. Also, the collector will be turned on for a longer period of time, which also increases the useful gain of heat.

39. The temperature at which the phase change occurs is 32°F (0°C). It is not feasible to operate a house at a temperature below freezing.

40. (a) Air   (g) Water
     (b) Water   (h) Air
     (c) Air   (i) Water
     (d) Air   (j) Air
     (e) Air   (k) Air
     (f) Water

41. (1) Blowing air across the hot storage tank and transferring that air to the house. (2) Heat exchanging between liquid storage and air in a secondary heat exchanger.

42. (1) Baseboard water heaters. (2) Radiant heating from water coils in the floor.

43. Not a good idea if the reflector and collector are left in a constant orientation. The reflector and collector positions should be interchanged between summer and winter operation.

44. R-23.

45. A variety of things, similar to the deductions allowed small businesses.

46. It obviously improves it tremendously because of the future leverage of ever-increasing energy costs.

Chapter 10

1. c.
2. d.
3. It has a much larger concentration ratio. This means that the solar energy density at the receiver is much higher and results in a higher operating temperature. Because of this, the second law operating efficiency is higher.
4. Double-axis tracking.
5. Because of the greater expense of the double-axis tracking mirrors for the solar tower approach.
6. The heliostats.
7. Single-axis.
8. Because of the need for considerable separation of the heliostats to avoid shadowing of one heliostat by adjacent heliostats. This is a particular concern to the fringe heliostats.
9. A tracking mirror.
10. (1) Biomass. (2) Solar tower facility. (3) OTEC. (4) Solar cells.

11. (1) By direct connection through transmission lines, (2) through storage into batteries or something similar, and (3) by using the electrical current generated to hydrolyze water into hydrogen and oxygen. These can then be converted back into electricity using a fuel cell.

12. The heat capacity of the water per unit volume would be exactly twice that of the rock per unit volume.

13. Advantages include light weight, use of inexpensive plastic and protection from the elements. Disadvantages include the transmission loss through the plastic dome and a shorter life span of the plastic mirrors.

14. The tension stress exerted along the fiber direction can be quite large before the fiber will break. This is in sharp contrast to the shear stress put on a conventional flywheel.

15. Energy in a more ordered form, such as electricity, mechanical motion, and so on. This is to be contrasted with heat stored at low temperatures in a fluid, for example.

16. Low-technology applications, such as flat-plate collectors all supply a warm fluid at low temperatures (less than 100°C). This is too low for the efficient generation of electrical power. Thus, these solar applications cannot reproduce themselves, that is, you need high-quality energy from another energy source to make the collectors in the first place.

17. Assume $\frac{1}{4}$ of all buildings can be fitted with solar space heating and $\frac{1}{2}$ with hot water devices. Also, the optimum size of a solar system is such as to provide about two-thirds of the total load. Thus,

$$\% \text{ Total} = (\tfrac{2}{3} \times \tfrac{1}{2} \times 8\%) + (\tfrac{2}{3} \times \tfrac{1}{2} \times 16\%) = 5.3\%.$$

Chapter 11

1. c.
2. b.
3. The conduction band.
4. The valence band.
5. In a conductor there are far many more electrons in the conduction band.
6. Because the energy of long-wavelength photons ($E = hf$) is too small to excite electrons across the energy gap.
7. A hole (or vacancy).
8. Their present high cost.
9. Silicon doped with an element containing five electrons (such as phosphorous). This increases the density of free electrons in the conduction band.
10. Holes.

11. (1) hf is less than, equal to, or greater than E_{GAP}. (2) For current to flow, $hf > E_{GAP}$ is required.
12. (1) The need for high-purity (expensive) starting material; (2) the slow crystal-growing process; (3) the large loss of silicon crystal because of the need to cut the silicon into wafers; and (4) the very labor-intensive fabrication process.
13. Yes, some, because of the thermal motion of crystal lattice.
14. Phosphorous.
15. Gallium (or perhaps aluminum).
16. To maximize the surface area.
17. Because the fraction of the solar spectrum that is effective in exciting electrons across the band gap decreases as E_{GAP} increases.
18. Yes. Otherwise the efficiency of the solar cell will decrease.
19. A factor of 2 comes from the average number of hours of exposure per day being about 24 in space (versus 12 on earth). The other factor of 4 comes from the ratio of extraterrestrial radiation to that holding at the earth's surface (on the average day).
20. Nothing to burn out, such as filaments in a vacuum tube. Also, the power dissipated is considerably less.
21. Much easier to form.
22. Reduces markedly the amount of expensive solar cell material needed.
23. (1) Only a portion of the solar spectrum is effective (that with $hf > E_{GAP}$). (2) When hf is greater than E_{GAP}, a portion of this energy is given up as kinetic energy to the electrons. (3) Junction losses. (4) The output voltage is less than E_{GAP}.
24. The region between the valence and conduction band. In principle no electrons are allowed in this energy region.
25. The hydrogen connects up with the dangling bonds, thereby decreasing the density of allowed energy states in the gap region.
26. 0.61 V.
27. 40 mA/cm^2.
28. 12.5 Ω for a 1-cm^2 cell.
29. 16 MA/cm^2.
30. Zero.
31. Zero.

index